MASTER THESIS

BEYOND BLOCKCHAIN

– Utilizing IOTA's Tangle technology to create a tamper-proof, immutable and scalable supply chain tracing concept to antagonize artisanal mining used in electronic vehicle supply chains

AKAD
UNIVERSITY

Examiner:	Dr. Thomas Fischer
Date of hand over:	18. 11. 2019
Name:	Julius, Bär
Matriculation number:	7976136
AKAD-Study Course:	Wirtschaftsingenieurwesen (M.Eng.)
Address:	Klosterbogen 16
E-Mail:	baer_julius@web.de

Neuried, 18.11.2019

Acknowledgements

At this point I would like to thank all those who supported and motivated me during the preparation of this master thesis.

My sincere thanks go to Infineon Technologies AG where I had a fulltime position during the entire degree. The flexible working hours and support of my supervisors were very helpful to the success of this thesis.

I thank my co-founders, with whom I started to build up localstoring.com intensely in the past months. Again, a flexible working style and cooperation was a key driver of success.

I thank the AKAD University for the new form of continuing professional education and the great support of the professors.

I thank my family and friends who supported me all the time, even in difficult phases. Especially my girlfriend, who always had an open ear for all my thoughts.

Finally, I would like to express my gratitude to my supervisor Dr. Thomas Fischer for the useful comments, remarks and engagement through the learning process of this master thesis. IOTA and electric vehicles are a big passion and interests of mine and it is not self-evident to be able to write a thesis on such new topics.

Julius Bär

Munich 18.11.2019

TABLE OF CONTENTS

LIST OF FIGURES

LIST OF TABLES

Abbreviations

DLT	Distributed Ledger Technology
Tesla	Tesla Inc.
CAPEX	Capital Expenditure
SHA	Secure Hash Algorithm
ICO	Initial Coin Offering
DAG	Directed Acyclic Graph
PoW	Proof of Work
QR	Quick Response
RFID	Radio-Frequency Identification
IATF	International Automotive Task Force
ISO	International Organization for Standardization
KCC	Kamoto Copper Company SARL
TNT	TNT Express B.V.
DHL	Deutsche Post DHL Group
BPMN	Business Process Model and Notation
UML	Unified Modeling Language
TWh	Terawatt hours

SCEM	Supply Chain Event Management
GDP	Gross Domestic Product

Formulas

$$H(B.N) < T \tag{1}$$

$$H(S) := SHA256(SHA256(S)) \tag{2}$$

1-INTRODUCTION

One company that has been on the watch list of tech investors for a long period of time and is controversially mentioned in the worldwide media nearly every day is Tesla Inc. (Tesla). The US-American manufacturer of electronic vehicles experienced strong demand growth of its cars and battery solutions (Tesla Inc., 2018a, p. 1) despite the negative news concerning their high cash burn rate (Tesla Inc., 2018b, p. 1). Tesla is often seen as the sole pioneer in the pure electronic vehicles segment, but in reality, they are only the signboard of this paradigm shift. The whole automotive branch is on the cusp of starting to become electrified. Almost all big automotive manufacturers and suppliers are investing in their electric vehicle fleet development, assembly plants, battery systems and secondary services such as charging stations and mobility solutions. The biggest German car manufacturer Volkswagen AG is aiming at 20-25% pure electric share of its total fleet until 2025 (De Bock, 2018, p. 5) while planning to spend 40 billion USD on development for electrified versions of current models until 2030 (Glencore, 2017, p. 15). In the US, besides Tesla, also more traditional manufacturers such as Ford are accelerating their plans to shift away from classic combustion engines (BBC News, 2019, p.1).

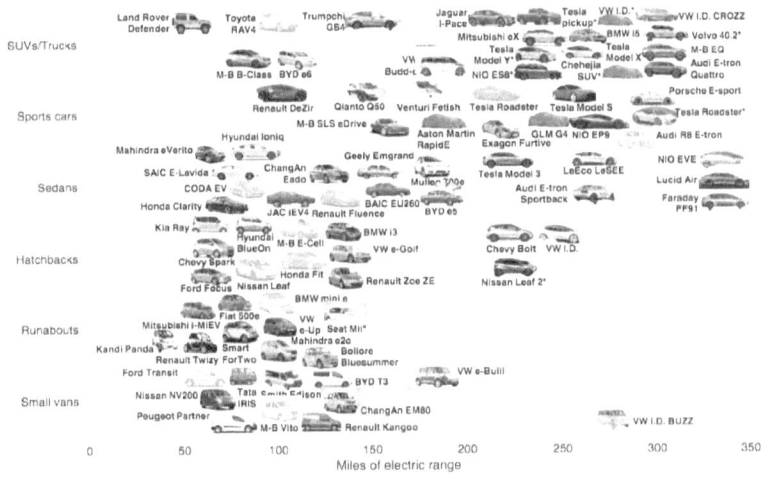

Figure 1: Models by style and range available through 2020
(Randall, 2017, p. 1)

But it is not only the OEMs who want to participate in this electronic revolution, even high-tech companies like Amazon Inc. are evaluating buying stakes in promising startups to participate in the fast-growing market (Handelsblatt, 2019, p. 1). It does seem that electric motors could replace fossil fuel engines in the next couple of decades. Although it has to be considered that the engineering of an electronic-based vehicle platform is requiring different materials to a high extent. It is expected that around three to four times the amount of copper and a multiple of cobalt are required to manufacture a pure electronic car in comparison to a car with a combustion engine (Tonks, 2018, p. 7). The majority of those raw materials are being used in the high-density lithium-ion battery packs who act as a power source for the electric motor. It is expected that each car will need 84kg of Copper and 8kg of Cobalt on average in 2030. In addition, the materials are used in secondary applications such as charging infrastructure or grid storage as well. Glencore, one of the world's biggest mining companies forecasts that cobalt demand in the year 2030 will be 314kt (kilotons) per year which represents a growth in demand of 332% based on the global supply from 2017. These numbers are calculated with a target market share of electronic vehicles of 30% (Glencore, 2017, p. 14).

1.1 Problem definition and outcome of the thesis

While lithium can be sourced worldwide through highly technical mining operations and with enough supply to cover the increasing electronic vehicle demand (Hocking, 2016, p. 4), supply problems could occur with cobalt and copper. The problem is that the supply may not be able to be catching up with the demand for various reasons. On the one hand, the copper market is underlying political risk due to its geographic occurrence mostly in unstable countries. On the other hand, dropping grades regarding the average amount of copper found in explored mining deposits are limiting the available supply, because it is simply not economical anymore for mine builders to extract the ore (West, 2011, p. 2 ff.). For cobalt, the available tonnage worldwide itself is very low, in addition, mining operations require high capital expenditure (CAPEX) and have long lead times from the first recovery to the final product. This indicates a supply deficit which cannot match the growing demand, ignited by the electronic vehicle paradigm shift (Marcelo, 2018, p. 1). The situation is also reflected in the price of the raw materials. Copper prices increased by over 50% in the past years and cobalt even tripled based on the levels of 2016.

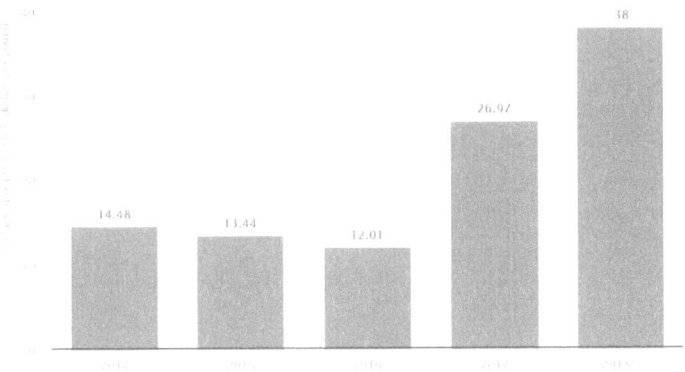

Figure 2: Average cobalt spot price in the United States from 2014 to 2018

(US Geological Survey, 2019, p. 50)

While only a high technological process can be used to mine lithium (Tran, 2015, p. 81 ff.), cobalt and copper can be found directly within the soil and therefore it does not require a plant to obtain smaller amounts of the raw materials. The possibility of simple and zero-cost production combined with high metal prices makes it very profitable for small miners to search for those minerals by hand. This so-called artisanal mining is very common especially in undeveloped countries with low economic status and a high poverty rate. Often people are forced into such kind of activates because otherwise, they would not be able to feed their families (Banchirigah, 2010, p. 158). Also, most of the mineral deposits are geographically positioned in poor countries such as Africa and some parts of Asia. The Democratic Republic of Congo (DRC) for instance produces over 65 % of the world's cobalt supply today whereof up to 20% is extracted by artisanal sources (Marcelo, 2018, p. 2). Artisanal mining in such areas is practiced under inhumane conditions, oftentimes even including child labor.

Due to being also harmful to the environment, small-scale mining is declared as illegal in many countries as well (Hentschel, 2003, 2. ff.). While car companies are trying to stop supporting those activates by relinquishing to source those minerals with artisanal origins it is not easy to ensure that they have bypassed them (Petroff, 2018, p. 1). Global supply chains today are complex and pass a variety of process steps beginning from the raw material to the final product.

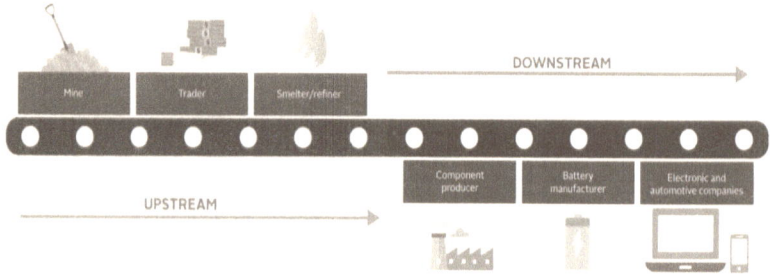

Figure 3: Top-level view of the cobalt supply chain (Dehman, 2018, p. 6)

Often times traders are consolidating previous input sources which makes it nearly impossible to track back the origin of a final component to its place of origin. This could mean that end customers buying an electronic vehicle cannot be sure that child labor or work with inhumane conditions was practiced to manufacture parts of their vehicle. In addition, the idea of electronic vehicles reducing pollution locally is useless, when even worse environmental impacts enter the supply chain across the globe at another stage of the product life cycle (Brennan, 2016, p. 3 ff.).

Simultaneously another megatrend which is mentioned in the media a lot of times in the past couple of years is distributed ledger technologies (DLTs), mostly known due to the term blockchain and the cryptocurrency Bitcoin. Blockchain-based technologies often promise to solve problems of data transparency while being tamper-proof (Rosenberger, 2018, p. 2 ff.). One technology which is standing out is the IOTA Tangle. The Tangle has a different approach to most known blockchain architectures and claims to solve the problems they face. In addition, one development scope and targeted use case the IOTA team is pursuing is the concept of a traceable, trusted and immutable method for supply chain traceability (Schiener, 2018, p. 1).

The question arises if the IOTA Tangle could solve the problem of electronic vehicle supply chains, while being cost-effective and scalable at the same time, suiting the requirements of today's supply chains. Therefore, the following research questions are addressed:

- What potentials do Distributed Ledger Technologies (DLT) have and what is the reason for that?
- How does the IOTAs Tangle differ from blockchain solutions, what makes its design unique and what are the advantages?
- What could be an effective process for companies to trace electronic vehicle supply chains utilizing the IOTA Tangle to solve the problems of artisanal mining?

1.2 Structure of the thesis

This thesis will first provide background information from the current research literature on DLTs, with emphasis on blockchain and bitcoin in chapter 2.1. Details on architecture setup and the procedure of transactions will be provided in section 2.1.1 and 2.1.2. Another focus will lie on the consensus mechanism which is used to verify transactions and the drawbacks of such an architecture. In chapter 2.2 another type of DLT, the IOTA

Tangle, will be introduced. The Tangle will be defined, the concept will be explained, and their special consensus will be highlighted. Afterward, the main differences between blockchains and the Tangle will be outlined in chapter 2.2.4. A concept of a supply chain tracing solution based on the Tangle will be introduced in chapter 3 using the findings of the literature review from chapter 2.2. The theoretical created methodology will then be applied to a real-world use case in chapter 4. The use case has the ambition to solve the problem of supply chain opacity within the electronic vehicle supply chain focusing on artisanal mining for raw materials. The setup of an electronic vehicle supply chain with its critical process steps will be highlighted in chapter 4.1. Finally, an outlook on DLTs will be given, the results of the Tangle-based tracing solution will be summarized and the answers to the identified research questions from the introduction be will be provided.

2. Distributed ledger technologies (DLT)

The need for new solutions on how to store, transact and verify data came up because of recently released privacy laws and a general skepticism regarding data security. In addition, data scandals from big corporations harmed the trust given to data origin and credibility (Jentzsch, 2011, p. 20). As a reaction DLTs positioned themselves as the solution for all these problems. DLTs are not describing one specific technology, it is more a term for a superset of technologies. The idea is that centralized collected and controlled data contains critical points of failure. This data oligopoly needs to be dissolved to prevent risks, misusage and fraud. DLTs are trying to solve the issues by decentralizing data management. Therefore, not a single entity is able to control the entire network. If multiple parties in the network are storing a copy of the same information, it is impossible for someone to manipulate the data set. DLTs are differentiating mainly in the aspects of who controls the network, how the network is set up and how transactions are processed. If the control is completely decentralized the setup is called a permissionless ledger, if it is distributed amongst many trusted parties it called a permissioned ledger (Chand, 2018, p. 647). Regarding the network structure DLTs use two new concepts besides the known centralized type. All types are shown in figure 4.

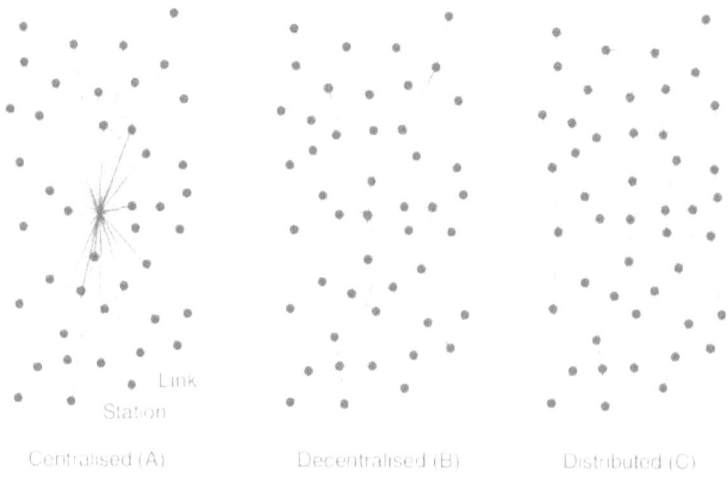

Figure 4: Centralized vs Decentralized vs Distributed Networks
(Truong, 2016, p. 4)

As stated earlier, the centralized network has a central owner who is in full control of the network. He represents a single point of contact who aggregates and shares all information flows. Therefore, this turnover knot represents a single point of failure which makes centralized setups very vulnerable. In this setup, the owner is also the only participant in the system who owns a copy of the data. Decentralized networks instead include multiple central owner points, each handing a copy of the stored data in hand. This balances the problem with centralized networks because if a single node is malfunctioning, all other nodes can still be accessed to attain the data. The decentralized network can be enhanced even further. If every node in the network has equal access to the data a distributed network is created. In this scenario, there is not a single point of centralization present. The cryptocurrency Bitcoin is set up on a blockchain architecture, which cannot be modified by a single entity and is, therefore, a decentralized network. At the same time, it uses computers around the world to set up its peer communication system which means that it contains a distributed network. This implies that both structures can be used at the same time for different objectives. In the case of bitcoin, the level of control is represented by a decentralized network, while the distributed aspect concerns the location setup of computers in the network (Truong, 2016, p. 5). Additional distinguishing features between centralized, decentralized and distributed networks are shown in table 1.

Property/behaviour	Centralized	Decentralized	Distributed
Points of failure	Single point of failure	Finite number of failures	Infinite
Maintenance	Easy	Moderate	Difficult
Stability	Highly unstable	Recovery possible	Very Stable
Scalability/ Max population	low scalability	low scalability	Infinite
Ease of development/ creation	Less Complex	Moderate	More details needed
Evolution / Diversity	Slow/little	High	High

Table 1: Structure differentiation by behavior and property types (Truong, 2016, p. 5)

To summarize, there are various types of DLTs existing and new network variants enter the market every day. This thesis will focus especially on the DLT of blockchain and the IOTA Tangle. The blockchain and its cryptocurrency Bitcoin are chosen because they are the most popular technology and they pioneered the idea of decentralization. IOTAs Tangle is chosen because it is already revolutionizing the idea of blockchains. With its unique setup, it is a one of a kind development project and in addition, supply chain tracing is one of their main roadmap goals. Therefore, the next chapter will introduce the concept of blockchain and bitcoins, in detail the setup, transaction processing and the used consensus algorithm. Afterward, the disadvantages of blockchains, that

IOTA's Tangle wants to overcome, will be highlighted.

2.1 Blockchain

The technology of blockchain and bitcoin was introduced in 2008 by a team or individual with the pseudonym Satoshi Nakamoto. The inventors or the individual is still not known. The idea was to transfer currency from one computer to another computer, more precise from one person to another. This transfer should happen directly from peer to peer, with no 3rd party control unit for instance banks involved (Rosenberger, 2018, p. 9).

2.1.1 Setup and transaction processing

The participants who interact with each other are individual users that are called nodes. Each user has assets assigned to his address, recorded in a so-called asset register. The asset register is a database that contains how many assets (e.g. Bitcoin) are assigned to each user. All participants save and manage this register and its whole transaction history of the blockchain in an invariable form on so-called nodes (Hein, 2019, p. 7 f.). Therefore, all transactions of the blockchain are publicly visible. Digital wallets are acting as virtual storage of assets holding the keys to get access to an account. Each user in the network has a unique address allotted to send or receive transactions. To secure the transactions a **public key method** is used. This means that every transaction has a public and a private key assigned (Burgwinkel, 2016, p. 7 ff.). The goal of this security feature is that only legitimate participants of the network can conduct transactions.

The **private key** is used as a digital signature, while the **public key** can be used to verify the signature by other users. Without the private key, it is not possible to do a transaction from a wallet. The public key is derived from the private key and while the public key is visible for each participant in the network, the private key is only meant to be seen by the user itself.

If user A wants to transfer value to another user, user B, A encrypts the transaction with the public key of the recipient, which is known to all participants in the network.

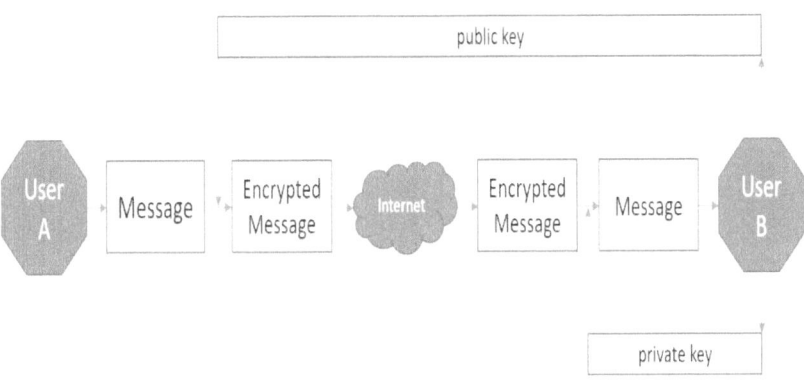

Figure 5: Transaction processing with the public key method
(Own representation)

User B is then able to decrypt the transaction by using his private key. To proof that the transaction is made by the sender and not another participant of the network, user A signs the transaction with its private key (Gayvoronskaya, 2018, p. 22 ff.). To create signatures a so-called hash function is used. A hash function is a one-way function which means the calculation is only possible in one direction. For instance, a hash function is used to create number Y from number X, but it is not possible to calculate number Y backwards to get X. The hash function changes a set of data with different lengths into an alphanumerical value with a fixed length. This method allows being able to identify a message without disclosing the content. The most popular Hash algorithm used in blockchains is the SHA-256 (Secure Hash Algorithm) (Eylert, 2012, p. 84 ff.). To sum up, in blockchains digital signatures are used to identify initiators or transactions and affiliation of value (e.g. Bitcoins) to a user's account.

2.1.2 Consensus algorithm

In general, a consensus mechanism is an agreement on the state of a database by the majority or in some cases all of a network's validators. It utilizes procedures and rules with the goal of maintaining a coherent state of facts shared between multiple nodes. For instance, in bitcoin's blockchain, the longest chain, therefore the chain with the most Proof of Work (PoW) done, acts as the valid ledger (Swanson, 2015, p. 4 f.). PoW is describing the consensus method the bitcoin blockchain uses. It is a mathematical puzzle that needs to be solved by so-called miners within the network. Miners are basically nodes that are validating and selecting transactions and submit proof of work to other nodes within the network. Those tasks have been split over the past years due to further development in the bitcoin network. Miners are still doing the selection of transactions and validating them but solving the proof-of-work puzzles is now be done by hashers. They aggregate recent transactions and use them to create a package called a block. The block is only valid if all contained transactions are validated. In other words, the hash of a previous block that was marked valid must be included. The hash links two blocks and is included in the successor and predecessor block.

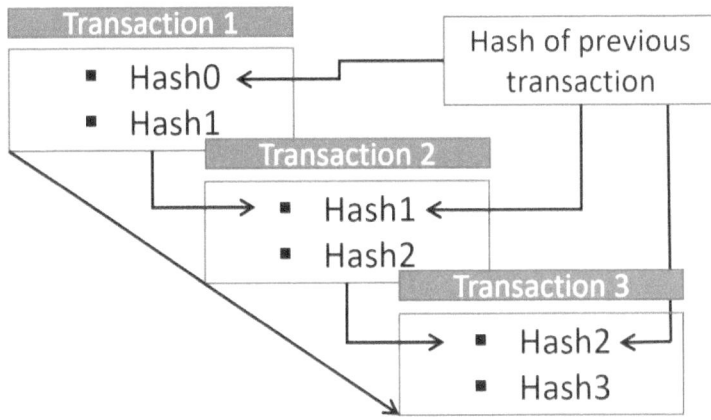

Figure 6: Hash dispersion in subsequent blocks (Own
representation)

The job of hashers is to search for a 32-bit number called a
NONCE (N) that results in the correct hash. They will randomly or
systematically choose values for N until the equation below is
solved.

Formula: H (B.N) <T   (1)   (Malone, 2014, p. 280 ff.)

"T" is a current target value that is recalculated by the network,
while "B" is representing the recent transactions. H is representing
the bitcoin hash function. As mentioned before, most of the time
SHA-256 is used.

H(S) := SHA256(SHA256(S)) (2)   (Malone, 2014, p. 280
ff.).

Blocks can only be created by trying to find the hash of the
previous block until one node succeeds and afterwards is rewarded
with bitcoins for its effort. Directly after that, the new block
becomes the "latest" block and acts as a starting basis to mine a
new, the next, block. This process is repeated every 10 minutes
(Buterin, 2014, p. 1 ff.). To summarize, the approval process of

blockchain transactions is stated below:

1. Transactions are enumerated in a block
2. Miners/Hashers are validating if transactions are legit by utilizing a PoW consensus
3. The first Miner/Hasher who solves the task and finds out the previous hash is rewarded
4. The validated transactions are attached as a block to the blockchain and step 1 starts again

2.1.3 Limitations and disadvantages

While the PoW consensus and the public key method of bitcoin and blockchain brings a lot of advantages, recent developments challenge its right to exist. The bitcoin protocol recently was developed further and even forked into new blockchain variants, but in this thesis, the traditional blockchain model for bitcoin (BTC) will be analyzed.

Transaction cost

Transferring bitcoins is not free. Since one block can only contain around 2000 transactions, in times where a lot of users want to do transactions there is a long queue. As long as no new block is found, all open transactions become pending. Miners, therefore, need to prioritize which transactions they want to include in the current block. What users are doing is attaching a fee to their transaction, providing incentives and making it more profitable for a miner to prioritize their transactions. People are basically out-bidding other peoples' fees to improve their chances to be selected to be in the current block. As shown in figure 7 the fee peaks in times where the bitcoin network gets used by a lot of users. In January 2018 the average fee for one bitcoin transaction was over 40 USD.

Figure 7: Average price per transaction (Coinmetrics, 2019, p. 1)

This is also one reason why bitcoin is entitled as digital gold. Due to the high fees, it should not be used for frequent and small transactions, more for storing asset value (Uddin, 2018, p. 6).

Transaction time

If the network is used by a lot of users at the same time, the queue is not only detrimental to the high fees, it also increases the transaction times. While the goal of bitcoins blockchain is to create a new block every 10 minutes, it can take a lot more time as shown in figure 8. This is because there are only a finite number of miners to process each block and there are a finite number of transactions that can be included in a block.

Figure 8: Average confirmation time for a bitcoin transaction in minutes (Blockchain.com, 2019a, p. 1)

Calculation and energy effort

As stated earlier the miners get a reward for solving hash puzzles and therefore for the creation of new blocks. But in order to find a hash, a lot of computation power is needed. While in the early days of bitcoin a standard computer was enough to "mine" bitcoin, today only giant computing centers with dedicated hardware are used for the hashing process. Computing requires a high amount of electricity besides the high CAPEX needed to set up a profitable mining solution (Schrader, 2018, p. 1). The current annual electricity consumption of the whole bitcoin network is 62 TWh (terawatt hours) which is equal to the power consumption of the country Ireland.

Another problem is that if more blocks get attached to the chain, the more data needs to be stored on all nodes. As shown in figure 9 the size of the blockchain was already over 210 Gigabytes in the first quarter of 2019. This transaction history has to be saved every time and uses a lot of storage space.

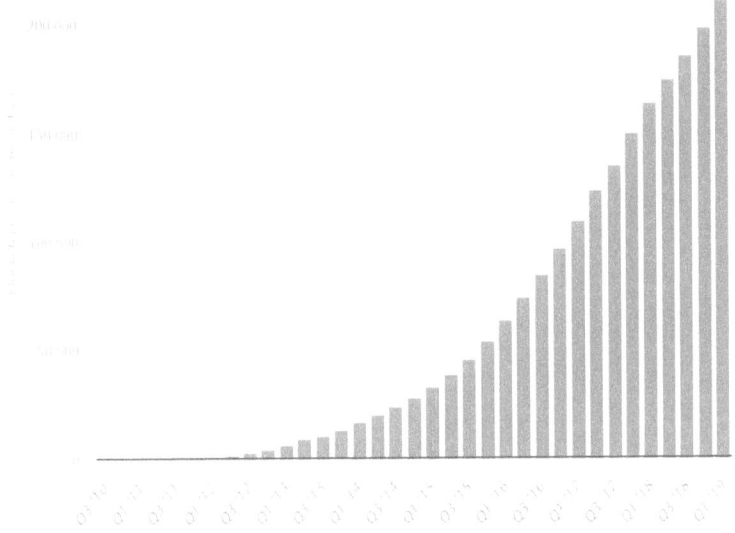

Figure 9: Size of the Bitcoin blockchain from 2010 to 2019, by quarter (in megabytes) (Blockchain, 2019, p. 1).

The energy cost and expensive hardware shows that it is only profitable for commercial so-called mining farms (computation centers) with specialized hardware to participate in the bitcoin network as hashers (Malone, 2014, p. 280 ff.).

Miner pools insecurity

Those miners often aggregate into **pools** to benefit from sharing the risk factor of never finding a hash. However, they will have to share the profits if one member of the pool is succeeding in finding the hash value. The context of mining pools within the bitcoin network is shown in figure 10.

All miners within one pool use the private key of the pool manager, who in case of success gets the reward and distributes it proportional to the computing power contributed by individual users.

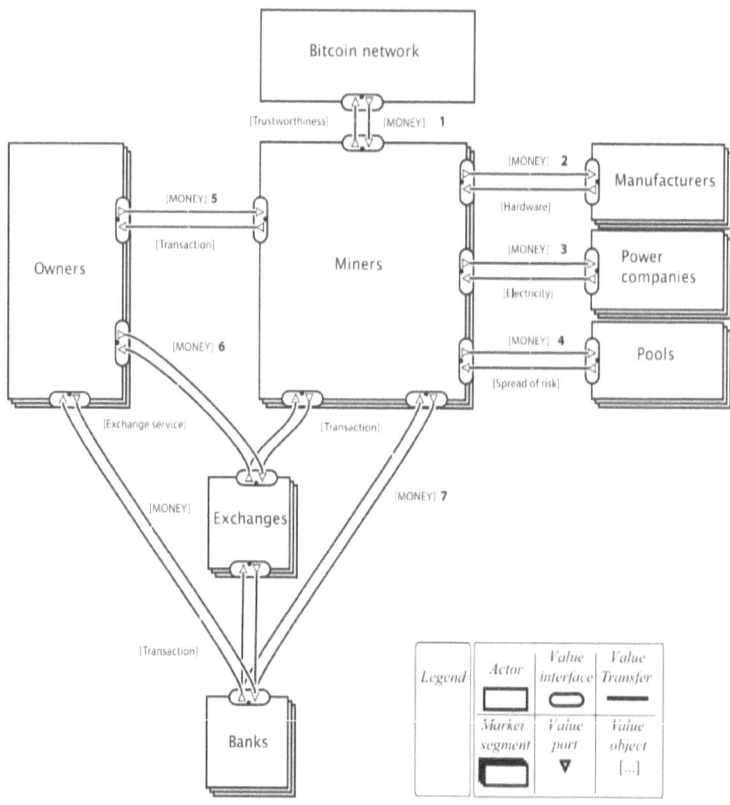

Figure 10: Value flows in the bitcoin network (Derks, 2018, p. 325)

The problem is that if pools get too big of a hash rate share, or if they join forces with another big pool they have options to harm the bitcoin network. As shown on the pie chart in figure 11, a few mining pools together would be enough to hold over 50% of the network's decision power. There are multiple options for how miners can harm the bitcoin network with their hashing power predominance. Selfish miners try to increase their incentive by reducing the probability of other miners to find a hash. At the same time, they want a revenue larger than their ratio of mining power. They do this by forcing the other miners within a pool to spend computation on wasted blocks which will never include a hash, which they have been working on before (Eyal, 2018, p. 95 f.).

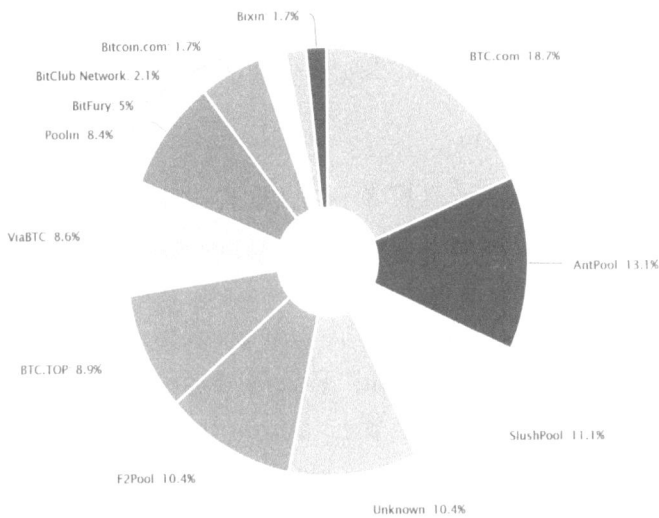

Figure 11: Estimation of hash rate distribution in 2019 (blockchain.com, 2019b, p. 1)

Another risk is mining cartels if they control over 50% of the hashing power, they could ignore blocks generated by other miners with are not within their cartel. They can even shut down the network if they decide to stop all hashing activities at the same time. Block withholding can be utilized even with less than 51% mining power, an individual withholds a found block from other network participants and mines it in secret. All those risks have the goal to increase the profits for individual miners at the expense of others, and mostly incentivize if greater centralization is applied. Contrary to the idea bitcoin seems not to be decentralized. The mining is very concentrated on a few mining pools that control the network (Courtois, 2014, p. 4 ff.).

The mentioned risks bring the need for a better protocol who solves the blockchains problems while keeping the advantages.

One development project who uses a different approach to blockchains is the IOTA Tangle. The Tangle claims to be the superior protocol to blockchains, solving all the problems with its newly created algorithm. The setup, transaction processing and consensus algorithm will be highlighted in the next chapter.

2.2 The IOTA Tangle

The cryptocurrency IOTA is a DLT which is based on a non-blockchain data structure. The main goals of development were to surpass the blockchain technology to solve the issues of scalability and high transaction fees. IOTA was created in 2015 by Dominik Schiener, Sergei Popov and David Sønstebø. The long-term objective of the token, the autonomous payment between connected IoT devices, was also responsible for the chosen naming (Popov, 2018, p. 1 ff.). IOTA hosted an Initial Coin Offering (ICO) from November until December 2015. On Dec 22, 2015, they raised 1337 Bitcoin, which was approximately worth 500.000 USD at this time, for project development. All tokens available were sold to ICO investors. There will be no new tokens created such as in bitcoin mining. The IOTA supply totals $(3^{33}-1)$ = 2,779,530,283,277,761 IOTAs = ~ 2,8 Peta IOTAs. In contrast, bitcoins can still be mined, and only around 85% of its total of around 21 million Bitcoin is available yet (Lie, 2017, p. 1).

The IOTA protocol is since 03.11.2017 managed by the IOTA Foundation which is located in Berlin, Germany. The goal of the foundation is the development, support and research on the IOTA-technology and DLT. The foundation has grown to a team with over 100 members since then and IOTA is ranked under the top 10 cryptocurrencies worldwide by market capitalization (Winheller, 2017, p. 1). The hype for IOTA was strongly driven by the market trend for the Internet of things (IoT).

2.2.1 Introduction and description of the system

The IoT market is expected to grow at a fast rate supported by the number of devices that are connected to the internet doubling from 2018 until 2030 (Strategy Analytics, 2019, p. 1). This new information technology area creates new opportunities for businesses but brings also new challenges with it. The three key impacts IoT will have are automation, integration and servitization. If devices are connected and are fully automated the need for fast and secure transfer of value between them will emerge (Lucero, 2016, p. 5 f.).

This leads to micropayments between IoT devices become more and more relevant. Since devices are owned by different entities on a global scale, the need for an industry standard for conducting data and value transactions is present – a topic classic blockchains such as bitcoin claim to solve. But for IoT devices the blockchain setup is not suitable.

On the one hand, paying a transaction fee higher than the actual value is not acceptable, on the other hand, it is hard to get rid of fees since they incentivize miners to create new blocks. A blockchain system is, therefore, always two- sided. The one side wanting to do transactions, and the other side which approves and validates them. This setup always creates certain discrimination of some participants. The disadvantages come from the setup and design of the bitcoin network and the need for a completely new system like IOTA arises. One of the main goals of IOTA is to become the industry standard for data security and transfer in the IoT space. One of their objectives is that anything with a connected chip inside can be leased and devices are able to trade resources with each other. This creates a Machine to Machine (M2M) economy which needs a suitable technology to be based on

(Schiener, 2017, p. 1).

The main difference to existing DLTs such as blockchain (bitcoin) is that IOTA is based on a **Directed Acyclic Graph (DAG)** consensus structure called the **Tangle**. Directed means that the graph is pointing to one direction, and acyclic means that the graph is non-circular. The Tangle is basically a sum of many vertexes which are representing transactions. They have two parents each which are representing the transaction they confirm. Similar to blockchains all transaction representatives are called nodes and are connected with each other. Each node stores at a given round a local DAG where each vertex, called site, represents a transaction (Quentin, 2018, p. 1 ff.)

The Tangle does not require miners to execute computational proof-of-work to validate sets of data (blocks). If a network participant wants to make a transaction, it has to perform a consensus beforehand, by validating two previous other transactions. Therefore, the network participants themselves become the "miners" (Tennant, 2017, p. 1 ff.). The Tangle graph acts as the ledger which is used for storing transactions. An example extract of the Tangle structure is shown in figure 15 and will be used as a reference to explain the setup and mechanism.

2.2.2 Setup and transaction processing

The Tangle is basically made up of **sites** and **nodes**. Transactions are getting handled by nodes, which issue and also validate transactions. Sites are transactions represented on the Tangle graph, they contain one or some transactions which relate together. A site can have three different statuses. Figure 12 shows the flow of the subsequent statuses a transaction can have. Fully confirmed transactions are marked green while the blue ones are only partially confirmed yet. Unconfirmed transactions, also called

tips are marked in red. The aim of every transaction is, therefore, to go from red to blue, to green. The green blocks are indirectly referenced by all the red blocks. Therefore, every confirmed transaction has a path leading to it from a tip.

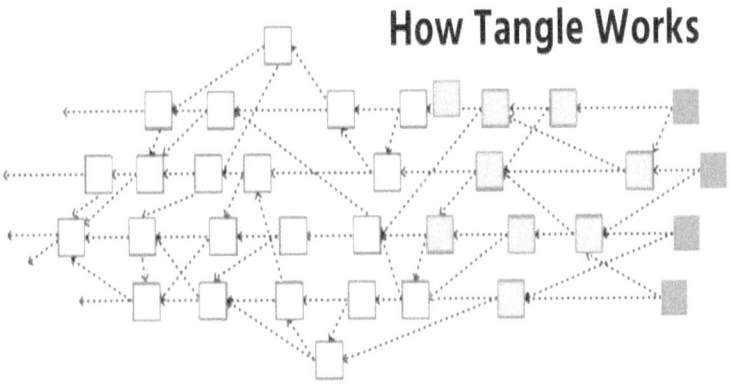

Figure 12: Transaction history by confirmation status (Maize, 2018, p. 1)

If a new transaction wants to be recorded on the Tangle the user's node (e.g. computer or phone) creates a transaction and signs it with its private key. Afterwards, the transaction must approve two previous transactions of other users. These approvals are represented by directed edges as shown in figure 12. There are two ways of how a transaction can approve another subsequent transaction. The first one is called a direct approval, where transaction A directly approves B (1). The second one is called an indirect approval which happens if some other transactions are getting placed in between A and B. In this case, A still indirectly approves B, even with C and D in between them. Both types of approval are illustrated in figure 13 below.

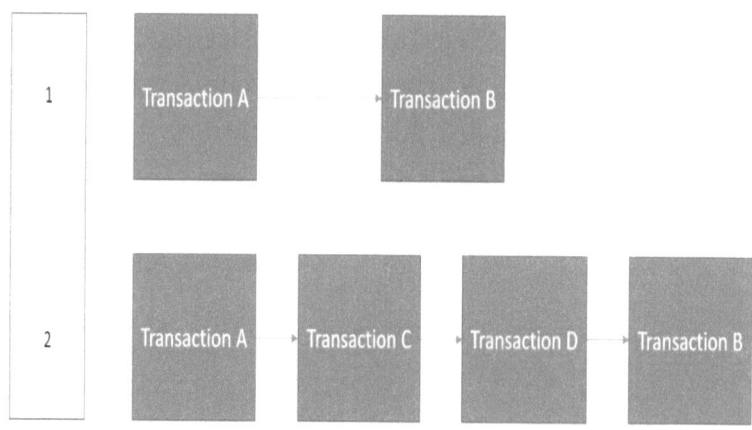

Figure 13: Direct and indirect approval of transactions (Own representation)

Therefore, a user who wants to issue a transaction is contributing to the security of the network, because he must work to approve other transactions first. In the beginning, a **genesis transaction** took place to set up the network. In the early days, there was also one address that held all of the previously explained ~ 2,8 Peta IOTAs. The first transaction made, the genesis transaction, split and sent all of the tokens to several other founder addresses (Gal, 2018, p. 1). The genesis transaction is approved by all other transactions as shown in figure 16. Every time transactions are contributed to the network, nodes check if the transaction is conflicting which means the nodes check the history of the Tangle and search for discrepancies. If found, the node will not approve the new transaction. If a node releases a transaction that approves conflicting transaction it risks that its own new transactions are not being verified by other nodes. The more approvals a transaction has, the higher the confidence of the system gets that the transaction is valid.

The approval step is similar to the one within the bitcoin blockchain. To finally issue a valid transaction, a node must solve a cryptographic puzzle known as the process of PoW to validate the two selected tips. The aim is to find a **nonce**, which is similar to the hash of bitcoin. IOTA uses a hash function called KECCAK-384 / KERL and a proof of work algorithm called Hashcash Lite. The addon light implicates that less computational power needs to be used to solve the puzzle, so all sorts of devices can utilize and contribute to the IOTA protocol (Elmaliah, 2017, p. 1). The aggregated process is shown in figure 14.

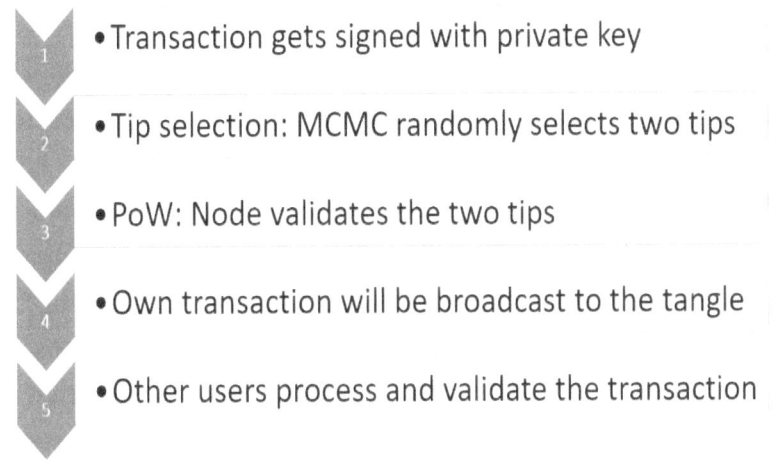

Figure 14: Process to publish a new transaction to the network (Own representation)

The set of transactions a node can choose its two transactions from is not always the same for every node. The IOTA network is acting asynchronous and different nodes can see different parts of the Tangle. Even conflicting transactions may be contained in the Tangle, but they will become orphaned due to no approvals given by nodes. Every node calculates a set of mathematical indicators, with one of them is how many transactions are received from other nodes. If a node is too "lazy" it will be excluded by its neighbors. This results in motivating the nodes to participate in the network even if they do not issue transactions at the time. The question arises on how the network measures a transactions overall approval. The nodes use a tip selection algorithm called Random Markov Chain Monte Carlo (MCMC) (Popov, 2018, p. 4 f.). A further technical explanation will be given in the next part of the thesis.

2.2.3 Attributes defining each transaction

Each transaction or transaction bundle (site) has specific attributes assigned to it, used to show off metrics such as the actual status and approval level to the network participants. Each transaction has a **weight** assigned to it. The weight is shown in figure 15 in the bottom left corner of each site. Therefore, the weight of site A is 1, the Weight of B is 3. The weight is representing the amount of work an issuing node invested into the transaction. It can be noted that transactions with a higher weight attached to it are in general more important than those with a smaller weight. An even more important metric of a transaction is the **cumulative weight** of a transaction. The equation of the cumulative weight equals the own weight of the transaction plus the sum of all weights of other transactions that directly or indirectly approve the said transaction. This is shown in figure 15, whereby the cumulative weight is the number in the top left corner of each transaction. Site F, for instance, gets directly approved by E and B, and indirectly approved by A and C. This results in the cumulative weight of F having the value 9 (own weight + sum of all weights of subsequent transactions). If a new transaction gets attached to the Tangle as seen in figure 15 on the bottom part, the cumulative weight changes (Popov, 2018, p. 4 f.).

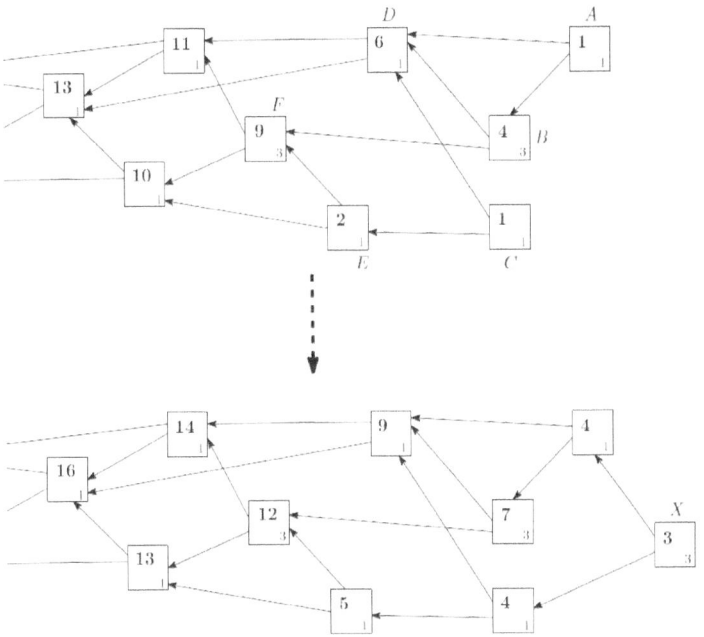

Figure 15: DAG with weight before and after a newly issued
transaction X (Popov, 2018, p. 6)

There are two more variables attached to a transaction which
are needed for the approval algorithm: **height** and **depth**. Height
refers to the longest distance between the genesis site and the
transaction. In figure 16 site G has a height of one. Depth
describes the exact opposite, the longest path from the transaction
to a tip. The example site G has, therefore, depth of four, because
of the longest path from F -> D -> B -> A, where A is the final tip.
The last metric that is part of the Tangle is the **score**. While the
cumulative weight was adding the weights of transactions to their
weight, the score is adding the own weight of previous
transactions to the own weight of the focused site. The score of C
in figure 16 is, therefore, seven, because of its weight of one plus
the own weights of E (1), D (1), F (3), and G (1) (Popov, 2018, p.
5).

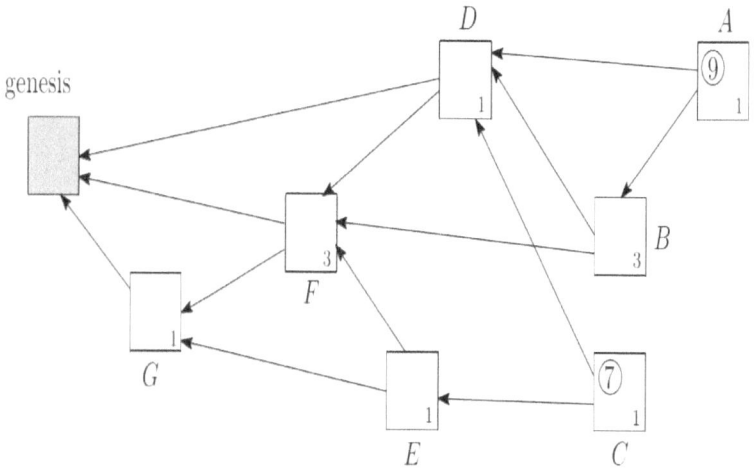

Figure 16: DAG with own weights for each site, scores
calculated for sites A and C (Popov, 2018, p. 7)

2.2.4 Consensus algorithm & the Coordinator

Based on all definitions and metrics that are available within
the Tangle network the next part will focus on how nodes select
tips to be able to confirm them. The IOTA Tangle uses a method
called **Random Markov Chain Monte Carlo (MCMC)** to select
transactions many times.

On the one hand, it is used to choose two other unconfirmed
transactions (tips) when creating a new transaction and on the
other hand to check if a transaction is confirmed. The level of
confirmation of a transaction can be assessed by running the
MCMC n times. The confirmation probability of a transaction is
represented by m of n, while m is standing for the number of
execution runs that resulted in a tip that has a path to the analyzed
transactions. This means that a transaction is 70% confirmed if the
MCMC gets initiated 100 times and 70 times it results in tips that
have a path to the transaction. A transaction is considered
validated if over 99% of runs of the MCMC algorithm result in a

tip directly or indirectly reach the transaction. Less than 50% would mean that the transaction is not yet validated and over 50% would mean that the transaction has a fair chance to be validated (Popov, 2018, p. 1 ff.).

To sum up the previous findings, a so-called MCMC algorithm is used to select transactions that are going to be validated. This happens based on certain steps that take into consideration the available metrics of each transaction such as weight, score, height and length. At this point, there is no need to research further on the selection topic and the MCMC algorithm because it is not clear how the future algorithm will work. Since IOTAs Tangle is currently in the development stage they created an interim solution to enable transaction approval instead of relying on the MCMC alone, while they are evaluating if this is the correct approach. The interim solution will be analyzed in the next chapter. As the research on the Tangle has shown, one of the main advantages of IOTA is the great scalability which creates trust if more and more users are contributing transactions. The problem is that in the current infancy stage while IOTA is still under development, a lack of a high number of nodes makes the network vulnerable for attacks by someone who owns a lot of computational power (GPU). The owner could easily create and approve malicious transactions.

This is why the IOTA foundation has introduced the **Coordinator**. The Coordinator is a set of special nodes run by IOTA to, directly and indirectly, approve transactions. This does not mean that the network is not decentralized. Even if a single node is manually controlled to steer the network, all other nodes are checking if the Coordinator is not breaking the consensus rules, e.g. by creating IOTAs out of nowhere. The Coordinator is basically a group of nodes that are stationed around the world. The nodes are run by the foundation and are forced to create zero value transactions every minute which directly or indirectly approves other transactions. The Coordinator's transactions are called **milestones** and are used to be referenced by standard nodes that are not run by the foundation.

Figure 17: Exemplary portrayal of the Coordinator issuing milestones

(IOTA Foundation, 2019, p. 1)

The goal is to be able to shut down the Coordinator at some point when the network itself is stable enough to carry itself (Quentin, 2018, p. 15). Another method to keep the network stable is the usage of **snapshots**. Snapshots are describing a process where multiple transactions that were leading to the same address, get grouped to one record where only the transactions with a value transfer are saved and the transaction history gets removed. The new address is then acting as the new genesis. This method helps to keep the Tangle history short and easy to handle, however, this process is currently done manually. In the development scope of the foundation is to make the snapshots automatically and so-called permanodes will be used to store the whole Tangle history permanent and secure (Rottmann, 2019, p. 1 f.).

The different approach of the setup and methodology used in the Tangle technology enables multiple advantages which will be summarized in the next chapter.

2.2.5 Advantages of the directed acyclic graph

The same metrics which were used to evaluate the usability and efficiency of bitcoin and blockchain technology will be used to analyze the Tangle. Therefore, the focus will lie on transaction cost, transaction time, energy consumption and security.

Transaction cost

Since no miners are needed for verifying transactions, no fee has to be paid to any party within the network. The incentive to conduct PoW within the Tangle is that only if a node does so, its transactions get approved. There is no fee to perform an IOTA transaction for all amounts of transferred IOTAs. This is especially useful in an IoT M2M economy where a lot of small value transactions are existing (Rosenberger, 2018, p. 50 ff.).

Transaction time

The transaction time is, in general, referring to the term "scalability". Scalability describes the ability of a system to react to growth to meet changing requirements without any loss of performance, stability and security (Weik, 2002, p. 1). In the case of blockchain and the Tangle, this can be translated to the question if the system is able to keep up to a rising load of transactions without changing its main parameters. Not using blocks as in blockchains makes the Tangle infinitely scalable. The more transactions that want to become verified, the more previous transactions have to be validated. This means that the more user the Tangle network has the faster it gets. This behavior can be described as hyperbolic.

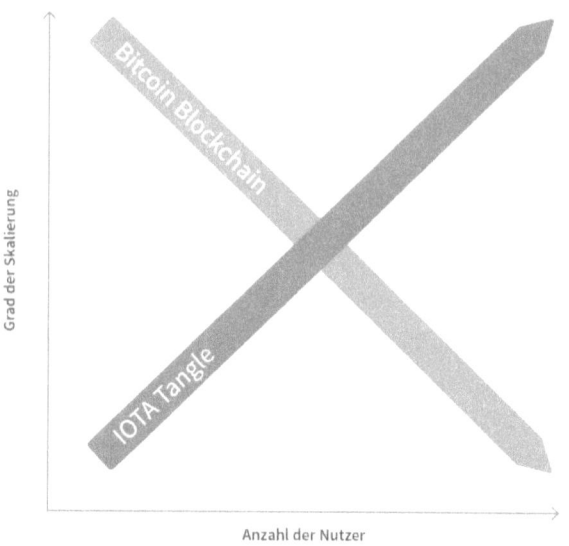

Figure 18: Blockchain vs Tangle scalability dependent on transaction throughput (Rosenberger, 2018, p. 52)

Energy consumption

Similar to the miners in blockchains, a PoW has to be conducted by nodes within the Tangle. The nodes use an individually for IOTA designed hash function called **Troika**. The goal is to create a non-hackable and energy-saving method to run PoW on nodes. Trinary algorithms, such as Troika, are more efficient than binary algorithms and therefore can adjust better to the growing performance of hardware. According to Moore's Law, the computing power of integrated circuits doubles every 12 to 24 months (Schaller, 1997, p. 52 ff.). As the IOTA foundation set the goal of every IoT device can become a contributing node such a hash function is necessary to satisfy the specific requirements. Besides the hash function, a change in hardware can also achieve reduced energy consumption.

To address this topic, the team behind IOTA initiated the project **JINN Labs** previously to IOTA. The currency IOTA and the Tangle are the software part complementary to the hardware JINN. JINN is a trinary processor which is similar to the Troika hash function not running binary, instead, it uses a trinary mathematical method. A trinary system uses three possible status (0, 1, -1) instead of binary, which only uses 2 (0, 1). As a result, the chips use a lot less energy and can compute faster. The JINN project is currently operated in secret by the foundation and there is little information available (Higer, 2019, p. 1 ff.).

However, the power of smaller IoT devices, e.g. a coffee machine, wearables or traffic lights will not be able to run a complete node even with their software and hardware adjusted. Since the goal is to let every IoT device contribute to the network, because the high number of devices combined still has a huge amount of unused computational power, the IOTA foundation introduced the concept of **swarm nodes**. The idea behind swarm nodes is to shard the core logic and database between multiple small nodes and let them collectively run it. This enables a cluster of devices to efficiently make a transaction without being a full node (Sønstebø, 2017, p. 1).

Security

The IOTA development team stated that all cryptography used to utilize the Tangle will be quantum-resistant. Quantum computers are expected to be a million or even more times faster than traditional computers. International Business Machines (IBM) is planning to release such a system already in 2019 with commercial solutions available shortly afterwards (Walters, 2019, p. 1).

The problem with the considerably more computational power is that it would be efficient to utilize trial and error methods to find a solution, for instance finding the nonce in the bitcoin blockchain. A quantum computer is said to be more than 17 billion times more efficient at mining bitcoin than the traditional computers the miners use. Within the IOTA protocol, the time needed to find a nonce is not much larger than the time needed to process other tasks relevant for issuing a transaction. Therefore, in comparison to the bitcoin blockchain, the Tangle is considerably more secure to attacks from quantum computers (Popov, 2018, p. 26).

To not secure the Tangle itself, but also the data which is transferred over the Tangle network the foundation has introduced **Masked Authenticated Messaging (MAM).** MAM enables sensors to encrypt and publish their generated data to the Tangle while being quantum-proof. Other parties who have the needed permission can then access those data streams (Sønstebø, 2017, p. 1).

Special features

Another feature of the Tangle that is also relevant for the next chapter of the thesis is the concept of **partitioning**. Due to the unique structure of the Tangle, it is possible to create a branch of the Tangle and later reattach it back to the network. Data integrity is still insured on the "side" Tangle even if the internet connection is temporarily lost. This is only possible because as shown in figure 19, previous transactions can reference multiple new ones. If the top and bottom side Tangles do not conflict, e.g. spend the same funds twice, they can simply merge later.

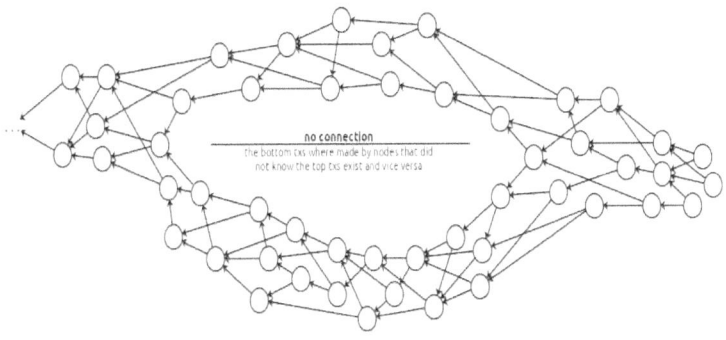

Figure 19: Re-merging of two Tangle paths after losing
connection (Helmar, 2018, p. 1)

This means that sensors located on a freighter ship can create an offline Tangle cluster while being on track on the ocean if they lose connectivity with the main Tangle. This is important for any global supply chain setup where remote areas without any internet connection available are the norm (Sønstebø, 2016, p. 1). In blockchains, a block can only be the reference to the previous one. Therefore, if two chains arise, they can never merge and the longest chain wins. This means that all transactions from the smaller chain have to be redone. This is also shown in figure 20, where the top row lost connection and becomes invalid after trying to reattach back to the bottom one.

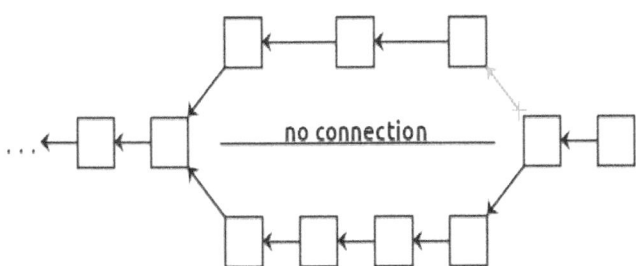

Figure 20: Upper blockchain becoming invalid after re-
merging with longer chain (Helmar, 2018, p. 1)

The idea of sub-Tangles is also an important metric of the goal to become infinitely scalable since full nodes will have a physical limit of processed transactions per second. There will be pools of nodes forming clusters who run their side Tangle because this is way faster than contributing all transactions to the main Tangle. Such a cluster can be within a company, a city or even a production site. Of course, a transaction can still reach any address worldwide by "hopping" between different clusters (Ivancheglo, 2018, p. 1). To sum up the previous chapter, the IOTA Tangle has many advantages over the blockchain protocol. In all shown aspects it appears to be the superior technology. A summary of the findings from the literature research is shown in table 2:

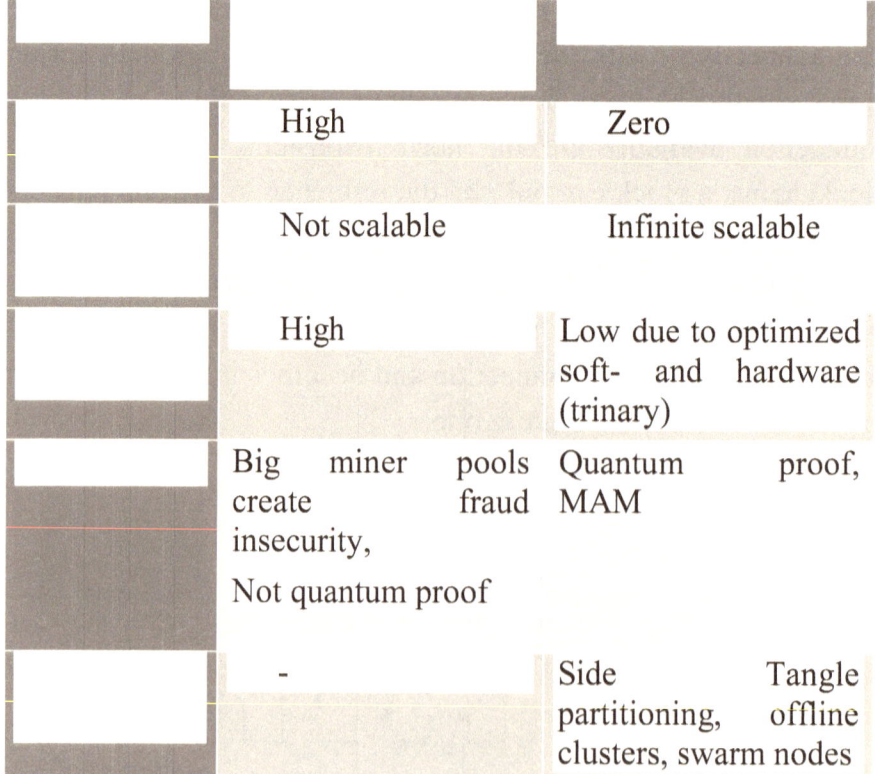

	High	Zero
	Not scalable	Infinite scalable
	High	Low due to optimized soft- and hardware (trinary)
	Big miner pools create fraud insecurity, Not quantum proof	Quantum proof, MAM
	-	Side Tangle partitioning, offline clusters, swarm nodes

Table 2: Comparison of Bitcoin and Blockchain vs Iota and the Tangle (Own representation)

The IOTA protocol running on the Tangle is not only better technology based on the research, but it also specifically addresses the idea of tracing global trade and supply chains. This leads to the Tangle being the base for the tracing concept which will be introduced in the next chapter of the thesis.

3. DEVELOPING A SCALABLE SUPPLY CHAIN TRACING CONCEPT BASED ON THE TANGLE

3.1 Concept development

In this chapter, the thesis will provide research and demonstration of an actual example of a supply chain tracing concept that is used in the industry. The main goal of this part is to ascertain what objects are needed to efficiently run and track a global trade supply chain which involves many complex steps. Afterwards certain points of failure will be highlighted with the main purpose of evaluating how they can be replaced with IOTAs Tangle. The last chapter, chapter 4, will then apply the methodical concept of a tracing solution utilizing the Tangle to solve the problem of artisanal mining within electronic vehicle supply chains. Further steps are summarized in figure 21.

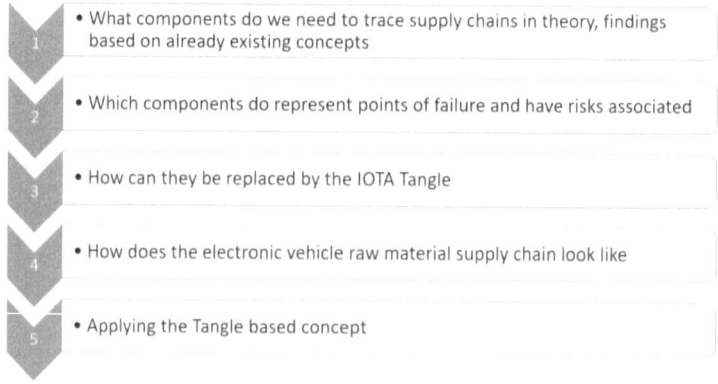

Figure 21: Methodology and steps needed to obtain the IOTA Tangle tracing concepts (Own representation)

3.2 Structure and network participants

A global supply chain consists of various **participants** contributing to the whole ecosystem. Participants include human beings, companies, technology, processes and products. Most of the times the actors are independently working in their separated environment and are not integrated along the complete chain. Since the main focus is to solve the problems of traceability, the area of technology and data is the most important. Participants are responsible for using the technology to record, forward and receive data. **Data** within supply chains especially used for tracing include timestamps, barcodes, labels, certifications, transport orders, bills of lading, pallets, loads, bank transfers and in general all records saved that are in the context of an end-to-end structure. It is important that this data is accurate, up to date as well as distributed and therefore, accessible for certain parties. On the one side, the technology should ensure those **security** attributes, on the other side the participant itself is also responsible for **proofing the origin** of the data. For instance, even if the data is correct, we do not know who provided this data (Leong, 2018, p. 6 ff.).

3.3 Procedure

One way of putting all sources together to get to a transaction history is the usage of Supply Chain Event Management (SCEM) systems. They work like a timeline that puts together all relevant events that were created along the chain combined with a time stamp. Here it is important to combine all events, means between different companies, different data categories and all kinds of sources. The goal is to get a rough overview when what happened in the supply chain and by whom. This information can then be used to optimize communication and material flows between parties (Wannenwetsch, 2010, 579 ff.). The records which get aggregated within the SCEMs can be created from different technologies. The best example are labels, mostly with a Quick Response (QR) Code attached. Those labels contain a basic set of data and can be accessed by scanning or manually reading the data printed on them. Because in global supply chains a lot of times multiple items are for instance in transportation on a big carrier ship, the need to access this information without seeing or touching the physical good emerged and therefore was developed later on.

One of the most used types of this approach is the utilization of **radio-frequency identification (RFID).** RFID chips can identify objects by getting physically attached to them. They store a small amount of information and provide them contactless to a reader. They do not require a battery source and are easy and cheap to manufacture. This concept was developed further by using Bluetooth frequency which enables greater reach but also needs a power source. This method is using so-called beacons. **Beacons** can be enhanced by adding sensors such as temperature, light or humidity sensors to provide more data to the recipient. Important for such a tracing solution is that the used hardware should be long-lasting and protected against for example dirt and water. Since packaging and tracing are mostly a non-value adding step for the end customer the cost should also be minimal. The best performance and cost structure is reached if different technologies get combined for different steps within the supply chain (Spengler, 2017, 576 f.). RFID and Beacon tracing concepts are easy to use: Every time a RFID scanner or Beacon reader takes note of a bypassing product, it gets recorded in the company's database. Therefore, a timeline and location can be assigned to a label that represents physical goods.

Every technology has its unique advantages and disadvantages.

For instance, Beacons have the best signal reach and amount of information that can be transmitted. However, a power source is needed, and the setup is more expensive (Gaudlitz, 2016, p. 1). Since technology varies in cost and functionality a qualitative overview of the different methods is shown in table 3.

	QR-Code / Labels	RFID (passive)	Beacons
Cost	Cheap	Medium prized	Expensive
Amount of data that can be transmitted	Small	Medium	High
Readable	Scanner	Gate	Beacons Tower
Power supply needed	No	No	Yes
Must be visible to read	Yes	Yes	No
Range	<1m	<4m	<30m

Table 3: Comparison of different tracing hardware solutions (Own representation)

Another important aspect is that, with the foresight to the electronic vehicle supply chain, some steps do have specific needs and therefore the use of each technology should be matched with its given characteristics. While a beacon setup is worth it for a bulk carrier which transports tons of cobalt, it may not be cost-effective for tracking parcels of single battery cells further down in the supply chain. A qualitative summary of this categorization is given in figure 22.

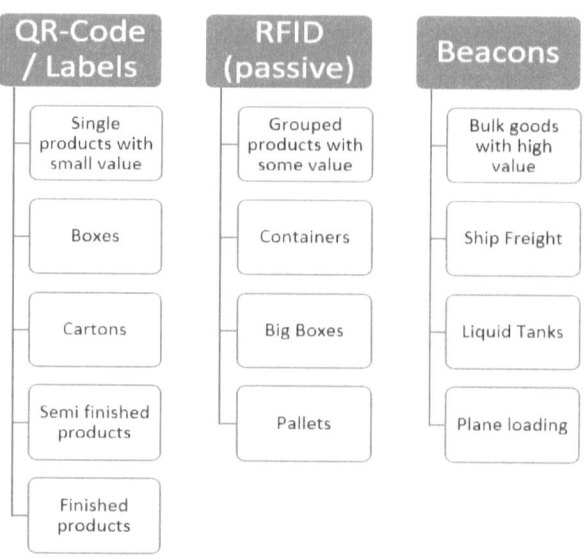

Figure 22: Matching logistical items to technology platform
(Own representation)

To sum up, existing tracing systems are made up of multiple technologies that collect data and store them within e.g. the cloud of a company. A major problem is that the single steps are not integrated from end-to-end and data breaches between steps lead to losing transparency (black box). Most of the data especially in low-cost countries where oftentimes the supply chain has its origin is paper-based and not standardized. There is also no flexibility given to adapt to changes, e.g. if products get repacked or the labeling is changing. With today's supply chains often including hundreds of steps, the historical records seem to be unusable further down towards the end customer. IOTAs Tangle can help to solve those issues, especially in **data**, **security** and **origin proofing**. While the data set will still contain the same kind of information, the Tangle ensures that it is accessible 24 hours, 7 days a week with no downtime. In addition, trough IOTAs security features the data cannot be changed afterwards or manipulated by a malicious party. Every transaction will also hold records of whom the transaction was signed and when. This is especially important to create a full and accurate history of the value flow. Integrating the Tangle within existing tracing systems is also possible and as stated earlier the best option is to combine different technologies. Beacons or RFID terminals need to be combined with trinary hardware to become nodes or even full nodes. They will then publish their generated data to the Tangle by issuing transactions containing information. As shown in chapter 2 there is no need to transfer value with such "data transactions". Therefore, it is easy and free for every participant to contribute their information to the Tangle network.

How existing systems have to be extended hardware and software-wise, to be able to connect to the Tangle is shown in table 4.

	Existing Supply Chain Tracing	Tracing via the Tangle
Hardware	RFID Terminal Beacons QR-Code (Handheld Scanner)	RFID Terminal Beacons QR-Code (Handheld Scanner) + Trinary Hardware
Software	Servers Cloud Computing	Servers Cloud Computing + IOTA Tangle

Table 4: Enhancing the existing tracing methods by the Tangle (Own representation)

In order to apply the concept demonstrated in this chapter, a real-world cobalt and copper supply chain will be shown in the next part of this thesis. Afterwards, the Tangle based concept will be applied directly to this example to demonstrated how it can be executed. This should act as guidance on how companies can better trace their existing supply chains by utilizing the Tangle.

4-Electronic Vehicle Supply Chains

4.1 Artisanal mining used for raw materials

As stated earlier artisanal mining is very common especially in undeveloped countries with low economic status and a high poverty rate. Families are forced into such kind of activates because otherwise, they would not be able to feed themselves (Banchirigah, 2010, p. 158). A rising cobalt and copper price is fueling this movement because other simple jobs become more unattractive from an income perspective. In addition, mineral deposits are geographically positioned in poorer countries such as Africa which is adding to the risks the population is exposed to even more. Global supply chains today are complex and pass a variety of process steps beginning from the raw material to the final product. In general, it can be noted that the closer the supply chain step is to the final product, the more valuable the goods and the higher-skilled the workforce needs to be. Therefore, component producers such as LG Chem who build pre-cursor materials, cathodes and assembles batteries, basically representing the whole downstream process, are located in developed countries such as the United States, China and Europe. OEMs are also investigating to integrate more of the electronic vehicle supply chain into their business model since the battery of a vehicle represents a big part of the value chain. A similar approach was shown with internal combustion engines, where OEMs tend to keep the development and production of the engine in-house while outsourcing nearly every other step to suppliers and contract manufacturers (Verband der Automobilindustrie, 2019, p. 1).

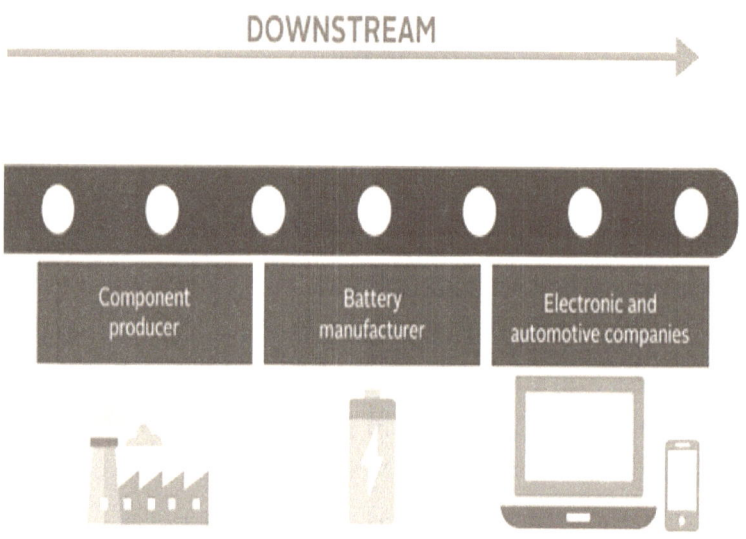

Figure 23: Downstream processes of the cobalt supply chain (Dehman, 2018, p. 6)

The German carmaker Volkswagen bought a stake of Northvolt AB, a battery manufacturer from Sweden at the same time, they are investing over a billion euros in an own factory for battery cells in Salzgitter (Heinz, 2019, p. 1). Simultaneously the big component producers and T1 suppliers such as Bosch, Continental or Infineon Technologies tend to be located in safe jurisdictions too, and even if they run operations in more unstable countries, they ensure the highest standards in their production and also in human labor conditions. As a result, the problems of unethically sourced and produced parts in an electric car mostly occur at the beginning of the process within the upstream processes. For this reason, the tracing concept and the next part of the thesis will especially focus on this area. It is not only the origin that is problematic in the upstream processes but at the same time, this is also the part of the supply chain where information about the origin and quality of material gets lost easily. Within the downstream processes companies mainly operate fully integrated, meaning they run databases, utilize cloud computing and trace their operations in the business warehouses. The upstream processes are in comparison non-transparent at all. This is mainly due to handwritten paperwork documentation or even further downwards the value chain, by private deals made between miners and traders.

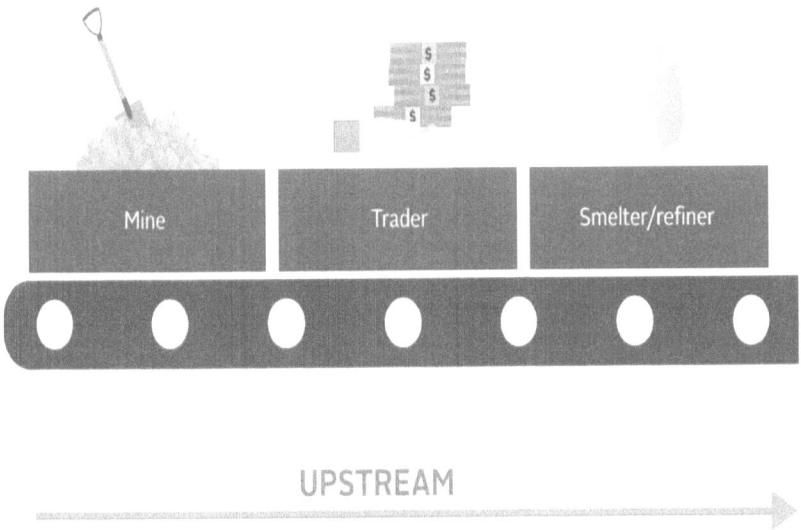

Figure 24: Upstream processes of the cobalt supply chain
(Dehman, 2018, p. 6)

In addition, traders are consolidating bought stock, selling it in bulk packages to the next party, which makes it nearly impossible to track back its origins. Every smaller campsite has a trader on-site which consolidates the mined ore and sells it to another bigger trader who is doing the same. The chain of different traders up to the final selling to smelters or refiners alone can be highly complex and non-transparent. Cobalt and Copper used in electronic vehicle batteries need high tech processing to suit the needs of the industry. Cobalt, for instance, gets processed by metal and chemical refineries to create cobalt sulphate, cobalt oxide and cobalt hydroxide. Since such treatments need expensive machinery and a highly skilled workforce, the bigger downstream processes tend to be located in more developed countries such as China, the US or Europe. Due to low labor and CAPEX, there are still a lot of local processors in the DRC which complicate the circumstances even more using harmful chemicals and unsafe workplaces. The DRC will be used as the start of the exemplary supply chain since it produces over 65 % of the world's cobalt supply and is often mentioned in the media due to its poor working conditions (Marcelo, 2018, p. 2). Artisanal mining in the DRC is practiced under inhumane conditions, oftentimes even including child labor. Due to being also harmful to the environment, small-scale mining is declared as illegal in many countries as well (Hentschel, 2003, 2. ff.).

In 2002 the DRC established a mining codex, which was revised multiple times in the past years. The codex is a set of laws released to attract foreign investors. As shown in figure 25 the investment made from overseas companies such as from one of the biggest mining companies Glencore plc was substantial. They bought and started to mine from huge areas of land where before artisanal miners were working on. Of course, the companies do not allow artisanal mining on their land due to reputation and economic reasons.

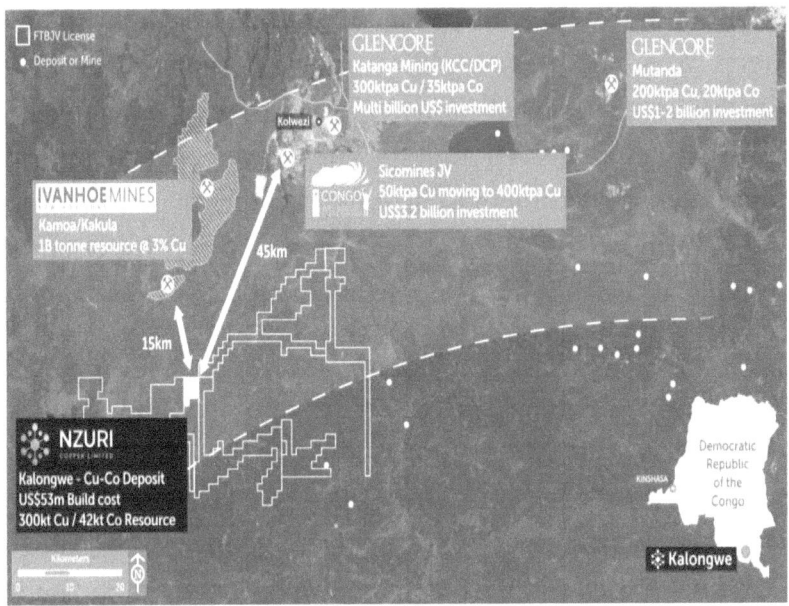

Figure 25: Kolwezi region - Ground Zero of the global battery metals boom US10-15 billion of new investment (Smits, 2018, p. 7)

The companies bought the land in the first place because of its high mineralization grades and ore reserves. Therefore, the artisanal miners also do not want to mine on other sites than the ones owned by the commercial big-scale operations. This leads to confrontations between companies and miners which often have a devastating outcome.

In June 2019 at least 19 miners were killed on a mining operations site called KOV because two galleries which secured an extraction area busted. The KOV mine is part of the Kamoto Copper Company SARL (KCC), one of the biggest cobalt and copper operations worldwide, which is owned by Glencore through its subsidiary Katanga Mining. In a later statement made by the company about the tragic incident, they noted that more and more illegal artisanal miners were extracting ore on their land.

After combing their land concessions with the help of the local military, they found over 2000 artisanal miners being on site. While Glencore urges all miners to not put their lives in danger by mining ore on industrial mining sites and helps to inform local communities, the problem of artisanal mining is getting bigger and bigger (Hume, 2019, p. 1). In contrast to these big scale mining operations, artisanal mining contributes between 0,5% and 2,4% of the whole Congolese Gross Domestic Product (GDP).

Companies do not want to support activities that include child labor, threatens the life of thousands and shortens life expectancy due to improper handling with chemicals. The only way to stop small-scale mining is to get traders to stop buying the product from such sources. This would lower the price and demand which results in miners have to search for other jobs. Another concept would be one step further, therefore smelters should avoid buying from such origins which include artisanally mined ore. This is even more important since many traders that are located on-site were previously miners themselves and stepped up the career. They still heavily rely on the income stream generated by reselling cobalt and copper mined by artisanal sources. Unfortunately, for processing companies such as smelters and refiners, traceability is virtually not possible at this stage. Even if they want to exclusively source from safe sources, the DRC is corrupt to a high extent and even state agents and union representatives label ores coming from illegal sites as ores originating from other legal artisanal mining zones (Tsurukawa, 2011, p. 2 ff.). It is therefore not only a problem of traceability but also of trust – two problems the IOTA Tangle wants to solve.

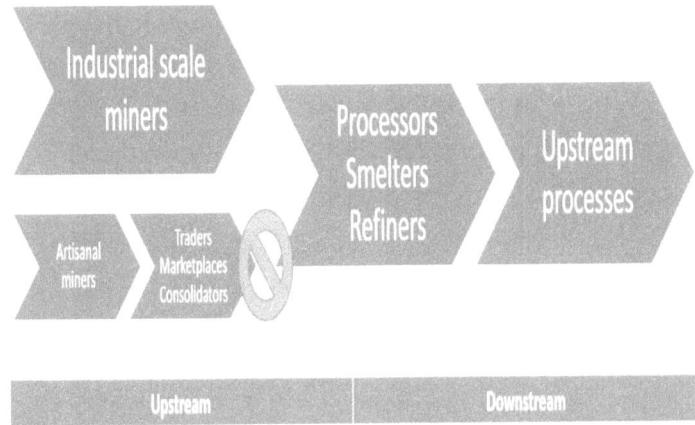

Figure 26: Prevention of unethically mined cobalt from entering the supply chain early on (Own representation)

Two points of failure can be extracted out of the previous findings. First, we have the point where artisanal miners sell their mined ore to the traders. The next step includes the traders buying such sources, combining them with legally mined ore and selling them further to the next party. The idea is that only industrial-scale mined cobalt and copper should enter the supply chain, eliminating the problem already in its infancy stage as shown in figure 26. However, it is hard to ensure that only the right material enters the supply chain at any stage with regard to the high corruption levels in the DRC. A smelter located in the DRC could still buy small scale mined ore because it is cheaper, and sell it labeled as legally sourced. Therefore, the processors themselves could become the problem too. In conclusion, the whole upstream processes need to be temper-proof, immutable and traceable from end to end. As shown earlier, the supply chain steps taken in the downstream processes are expected to be safe and trusted and will therefore not be weighted any further.

To analyze even in more detail, in this chapter a global demonstration of the cobalt and copper supply chain is given in figure 27. What can be noted is that the start of the supply chain is mostly in the DRC, while further processing happens to take place in mainland China. End customers or component producers tend to be more spread out globally to Europe and the USA as well as Asia. This shows the importance to solve the problems mentioned before, already at the beginning of the supply chain, since the processes and steps are more focused in one area with a small number of parties included. Also, as found out before, the second half of the electronic vehicle supply chain tends to be in more stable and secure jurisdictions where integrity and auditing by the government agency's takes place more often.

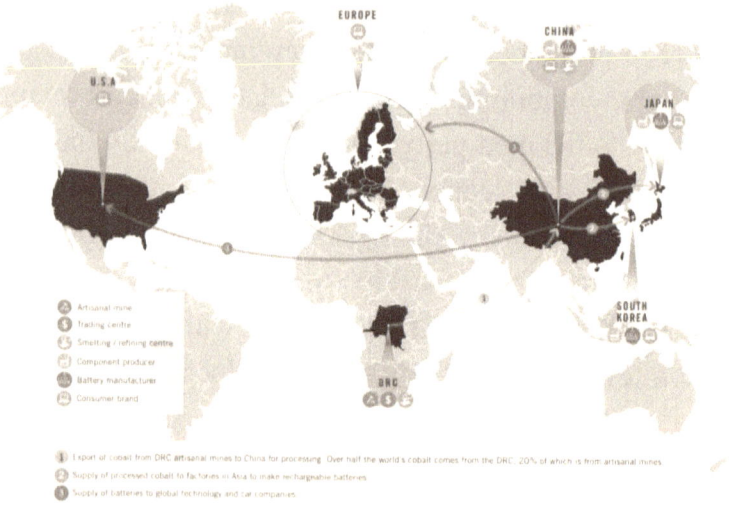

Figure 27: Movement of cobalt from artisanal mines in the DRC to the global market (Amnesty International Ltd, 2016, p. 46).

However, the global scale and broadness of the network is another topic which makes it extremely difficult to trace back origins. There are multiple types of transportation, unique packaging units and different documentation required to enter multiple countries. Since the sum of possible sources and supply chain setups is widely based, a single real-world example of an electronic vehicle supply chain is shown in figure 28 as a reference object for the Tangle-based tracing concept provided in the next chapter. This will demonstrate how a tracing concept can be matched and adjusted to fit a specific supply chain variant.

The supply chain starts with artisanally mined ore in the Katanga province in the southern DRC. Miners who reached a certain level of status begin to switch from mining to trading ore. The so-called pit-side buyers have limited capital and work with simple scales. They sell their products to the next bigger buyer which operates at a higher level. All these trading activities are largely untraceable and only the larger buyers with an entitled business in larger cities keep records of their operations.

The hand-mined ore gets consolidated multiple times by smaller traders until it reaches a licensed buying house or the open market. Licensed buying houses are often partly state-owned and hold a right to acquire big amounts of ore and are entitled to transport and sell it nationwide. Similar to non-licensed traders they often do not require an identification proof of ore origins. This results that the mixed bulk material of copper and cobalt could include child labor or artisanal origins (Vlassenroot, 2008, p. 36). Buying houses are located in medium-sized to large cities such as Kapata and Kamina in the middle of the Katanga region. An audit conducted by LG CHEM took pictures of children carrying bags of ore to those open markets but refused to investigate further since they had to leave the area due to security concerns (Ahn, 2018, p. 25 f.)

The transportation from the local traders from the south to the middle region is done by cars, trucks and in some cases trains. No logistical documents are available here and another step where ore of different origins is mixed up is given.

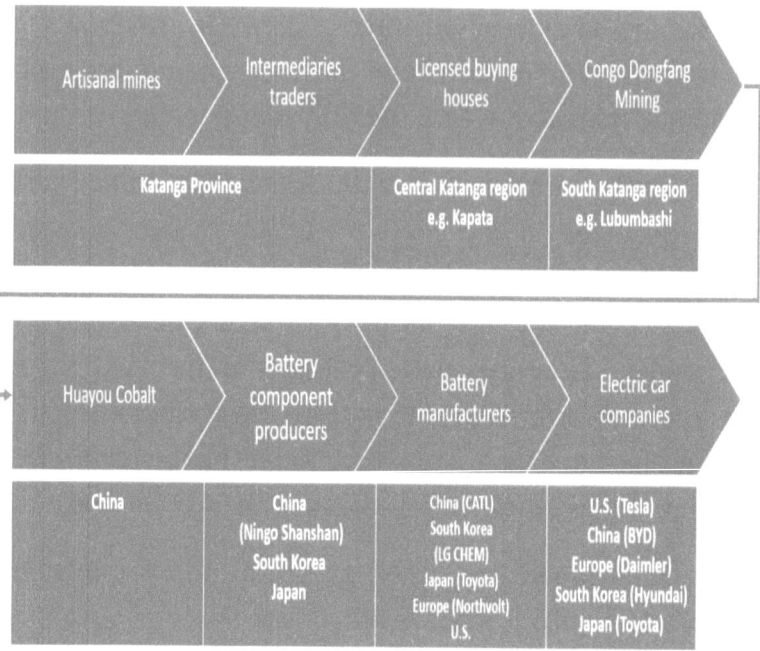

Figure 28: Possible real-world end to end supply chain of electric vehicle (Own representation)

The licensed buying houses are now selling their consolidated ore to Congo Dongfang Mining, a processor unit of Huayou Cobalt, a Chinese company specialized in the production and distribution of different types of cobalt. Huayou Cobalt (Huayou) is located in China but operates some of its processing plants directly in the DRC. Over 20% of Apple's cobalt required for their cellphone batteries is supplied by Huayou (Frankel, 2016, p. 1). Congo Dongfang Mining runs its operations in the capital of the Katanga Province, in a city called Lubumbashi which is in the

South West. After processing the ore at its industrial-scale plant, the company packs and transports the refined cobalt to china where the next processes take place at plants owned by the mother company Huayou. The company has its operations located in Zhejiang, a coastal province in eastern China. To get the cobalt to the facility the company purchases transportation services from a logistics company that brings the ore from Katanga to the port in Daressalam. To get to port, the train has to enter the border and pass through Tanzania. Since the amount of cobalt produced in comparison to copper is rather small, Congo Dongfang Mining shares the trains with other companies. This creates a new risk of mixing ore of different origins and also logistical documents required to enter customs at Tanzania and enter the port differ. It cannot be ensured that the information labeling the transported goods is accurate and trustful. After finishing processing and the first part of the downstream supply chain, Huayou ships the refined cobalt to battery component manufacturer Ningo Shanshan. The supply chain is at this stage spreading out to more customers, also located in Japan or South Korea. Here the supply chain is becoming more and more transparent, entering safer and better-controlled jurisdictions (Institute for Economics & Peace, 2019, p. 8).

Cobalt is used in Lithium-Ion batteries especially for the cathode material. One of the most used cathode materials is lithium-cobalt-oxide (Li-Co-O2). Cobalt is so important because it is one of the variables which are relevant to increase the energy density and therefore the range of an electronic vehicle (Smith, 2018, p. 1). After the cobalt is manufactured into cathodes, the products get shipped to Huayou's customers, which are battery manufacturers. Battery manufacturers are mostly located in Asia such as China's CATL, South Korea's LG Chem or even first producers from Europe such as Northvolt. The battery manufacturers often have supply agreements with major OEMs, directly supplying them with the demanded batteries. The automotive manufacturers are in a rush to secure supply deals from the market, fearing the free supply cannot keep up with their demand. For instance, the biggest car manufacturer by volume Toyota is sourcing its batteries directly from CATL, while Volkswagen is relying on Samsung SDI or their own factory with partner Northvolt (Whitley, 2019, p. 1).

The research on the example for a typical supply chain provided one important finding. Because of the high amounts of different steps needed in the end-to-end battery supply chain combined with multiple points of failure, it is nearly impossible to ensure legally mined ore in the final product if the current supply chain is kept intact. This is because of the high corruption in the DRC, the unregulated resource markets and the lack of proper documentation.

Based on this, the idea of figure 26 will be used as a strategy in the next part of the thesis. The methodology is not to trace back the batteries components to see if they result in artisanal mining operations, moreover, the concept developed on the Tangle should ensure that only industrial-scale produced material is entering the supply chain in the first place. This means the downstream processes including traders and trading houses need to be cut off the global supply chains. The approach seems more reasonable because big mines that are operated by major companies can prove the origins of their ore more easily. In addition, industrial-scale mines are not interested in artisanal sources because their productivity is already at a much higher level.

The bigger companies are aware that they can use the term of ethnically sourced cobalt as a selling argument to increase their margins and floor price. DRC based Nzuri Copper Ltd. states in their investor presentations that they especially conduct a corporate governance strategy contributing to anti-bribery policies and the local environment. They directly aim at the multi-national car companies who are undergoing increased pressure to identify on their own where their battery components come from (Smith, 2018, p. 13). Also, they do not want to risk losing their customers by harming social and environmental standards, such as Huayou did in 2016. The company was exposed by a report from NGO Amnesty International which found out that Huayou's subsidiary Congo Dongfang Mining bought cobalt on open markets from all kinds of vendors. The press demasked and denounced electronic companies such as Samsung, Apple and Tesla accusing them of not being aware of their full supply chain standards (Amnesty International Ltd, 2016, p. 8).

In order to apply the Tangle-based tracing concept, the upstream processes of figure 26 need to be split even more to see exactly where possible risk sections are.

Applying the supply chain tracing concept to EV supply chains

The supply chain can be divided further into three main sequences. First, we have the mining operations where the ore gets extracted from the ground and is further processed on-site. Those high scale mines are owned by multibillion-dollar resource companies due to the fact that a lot of CAPEX and mining expertise is needed to run a successful mine. The biggest copper and cobalt mines in the DRC include the Kamoto, KOV and Mutanda operations owned by Glencore or its subsidiaries and the Tenke Fungurume mine owned by Freeport-McMoRan and

Lundin Mining. State-owned Gécamines is oftentimes holding shares in copper and cobalt producing mines forming joint ventures since they bring political power, local knowledge and exploration jurisdictions with them (BGR, 2017, p. 7 ff.). If smaller exploration companies such as Nzuri Copper Inc. find ore by drilling test work, they often get instantly bought by one of the major conglomerates. Therefore, all operating mines are owned by a few key companies in this area.

All processes conducted by those major conglomerates beginning from mining, processing, stocking and packing of the ore are summarized into one system in figure 29 (system 1). All operations take place within the company's legal entity shell, on their land concession and through their employees. Due to the size of the operations the data infrastructure is running on stable, traceable servers who generate, save and distribute data of the daily activities. The working conditions are also more stable and environmental aspects get monitored carefully.

System 1 itself seems stable regarding its inner processes. The ore always stays on the same land, only allowed and verified workforce can enter the location and no one is incentivized to use artisanal sources. It is expected that the only thread to this step of the supply chain is the intrusion of external ore into the system. Marked as red 1 in figure 29 this risk can occur for example by artisanal miners entering the land such as in Katanga Mining's incident. Other risk factors include corrupt personal buying artisanal sources to feed into the industrial-scale processors or inventory to improve the company's performance to benefit their own goals.

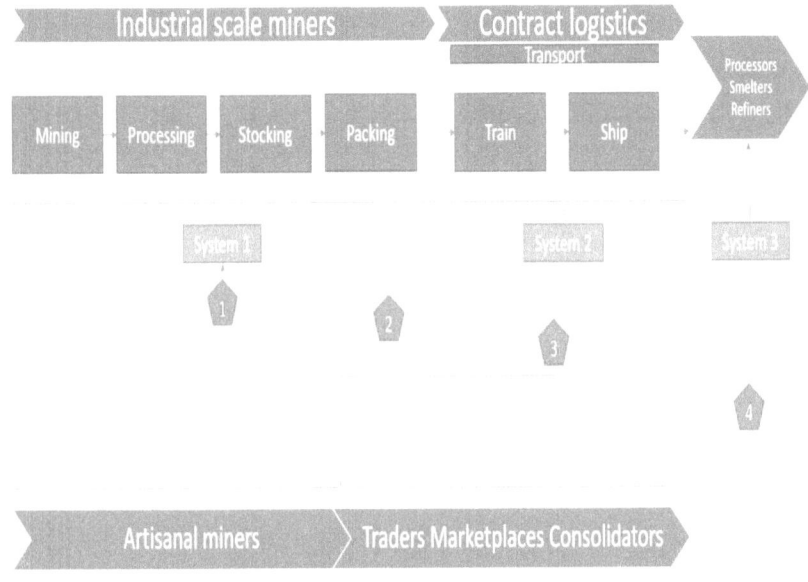

Figure 29: Points of failure due to artisanal infiltration in the downstream processes (Own representation)

System 2 is consisting of the contract suppliers who perform logistic services for the mining companies. They run their own part of the supply chain including different types of transportation methods and freight. They take the packaged goods from the mining site and bring them to nearby harbors by train or trucks. The train or trucks then get unloaded and the goods get loaded on heavy bulk ships that head to China for further processing. Since the transportation company is only responsible to move goods from point A to point B, they have no interest in what kind of goods they transport. As mentioned earlier one single mine is often not producing enough ore to fill an entire ship or train. Therefore, logistics providers often combine and aggregate loads from different sources. The risk, marked in red 2 and red 3 in figure 29, is that artisanally mined cobalt gets mixed with industrial mined ore and enters the supply chain. System 2 itself seems stable regarding its inner processes, such as system 1. Major logistics providers operating in the DRC such as TNT Express Group B.V. (TNT), Deutsche Post DHL Group (DHL) and Bolloré Logistics have an integrated IT infrastructure and certified fulltime employees working for them. Other risk factors include corrupt personal that infiltrates processed ore from artisanal sources into the same freight load.

System 3 is made up of the processors, smelters and refiners who can still be located in the DRC or partly in China. If only industrial-scale mined copper and cobalt gets into their operations, the chance is high that no artisanal source will enter the supply chain until the end product, the electronic vehicle. This is due to the fact that afterwards the downstream processes are located in safe jurisdictions with higher transparency standards. The risk factor, marked as red 4 in figure 29 is, therefore, the chance that processors directly buy artisanal sources due to the lower price. It has to be ensured here that they are only getting industrial mined ore for further processing. The unethically sources can enter the supply chain on the one side trough corrupt employees and businesses and on the other hand by falsely labeled ore from industrial mines. The processors have to be sure that the industrial-scale mined copper and cobalt is coming from their mines and are not just labeled falsely.

All risks from the analysis above will be separated and examined in more detail in the next part. The Tangle-based solution will be used to address the four extracted risk factors.

Risk factor 1: Artisanally mined ore enters the industrial-scale operations

Possible reasons:

- Artisanal miners operating on the company's land

- Artisanally mined ore gets bought by the company to feed it into their processing equipment
- Artisanally mined ore is labeled as it would have been produced by the company

A solution to reason 1:

Every mine site should be surrounded by fences, due to high aggression and violence, some companies even hire private armed security contractors to secure their land concession. Chinese owned Tenke Fungurume Mining one of the largest cobalt und copper producers from the DRC asked the local military for help to clear their land from illegal miners. Over 10.000 artisanal miners were removed voluntarily by a battalion of around 700 Congolese troops in June 2019 (Agence France-Presse, 2019, p. 1).

A solution to reason 2 and 3:

To be able to ensure that only the industrial mined ore is entering the companies processing steps, the idea is to give every item mined by the company its unique identifier. This identifier needs to be trustful and immutable, so no one is able to change it afterwards. This is a fairly new concept since mining companies do trace its stocked ore but not on a small granularity level.

Mining companies receive mining exploration permits from the government. They basically allow the company to mine ore in this area. For instance, Kamoto Copper Company, a subsidiary of before mentioned Katanga Mining Inc. (Glencore) owns the following exploration rights:

Property	Exploitation Permit Number	Rights Granted	Area of Title	Valid Until
KTO UG and Mashamba East OP	PE525	Cu, Co and associated minerals and metals	13 blocks, 11.04 km^2	April 3, 2024 Renewable
T17 OP and T17 UG	PE11602	Cu, Co, Ni and Au	2 blocks, 1.699 km^2	April 3, 2024 Renewable
Extension of Kananga	PE11601	Cu, Co, Ni and Au	1 block, 0.849 km^2	May 7, 2022 Renewable
KOV OP and KOV UG	PE4961	Cu, Co	10 blocks, 8.49 km^2	April 3, 2024 Renewable
Tilwezembe OP	PE4963	Cu, Co	9 blocks, 7.64 km^2	April 3, 2024 Renewable
Kananga mine	PE4960	Cu, Co	13 blocks, 11.04 km^2	April 3, 2024 Renewable

Table 5: KCC's permits and legal tenures in regard to concession areas (Kamoto Copper Company S.A. (KCC), 2018, p. 40)

Each permit covers a specific area coverage in the DRC. For example, exploration permit number PE11601 allows Katanga Mining Inc. to mine Copper (Cu), Cobalt (Co), Nickel (Ni) and Gold (Au) on an area which is 0.849 km^2.

Figure 30: Mining permits held by KCC (Kamoto Copper Company S.A. (KCC), 2018, p. 47)

Since this area is still too big, the need for a smaller granularity identifier arises. Another metric that is used by mining companies to specify ore grades on drill hole level is to assign a name to each drill hole. For instance, previously mentioned Nzuri Copper has found 1,85% Copper mineralization at their drill hole "DKAL_DD124" (Smiths, 2018, p. 20).

If we now combine the legal right concession with the drill hole naming and add a counter, we get a number that enables us to locate ore to its exact origin. An example is illustrated below:

Number of exploration/mining permit + Name of drill hole + counter = unique origin identifier

-> PE11602 + DKAL_DD124 + 0001 = PE11602DKAL_DD124_0001

The mining team and their IT system should now automatically create a unique number X every time they mined a specific amount of tonnage or duration. This number X should be attached physically to the unit by sticking a QR code on the carrier (e.g. full truck or box). At the same time, the system who created the number X acts as a full node by creating and sending a hash (SHA256) to the Tangle with the information of the origin (e.g. PE11602DKAL_DD124_0001) included. The hash is immutable and has a timestamp assigned. The Tangle then verifies this transaction by running the consensus algorithm. The digital identity can now be scanned by all parties e.g. with a smartphone or computer and all relevant information is being displayed.

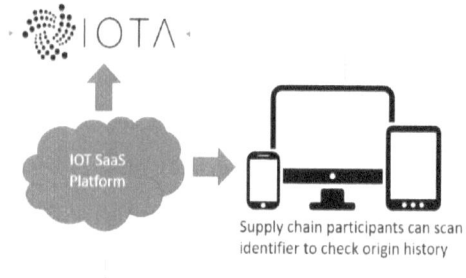

Supply chain participants can scan
identifier to check origin history

| Industrial scale miners | Contract logistics | Processors, smelters, refiners |

Figure 31: Adding an IOT SaaS Platform in the tracing setup
(Own representation)

It is recommended to add an additional participant to the setup to make the process of publishing and receiving information easier. In the setup shown in figure 31, an IoT Software as a Service (SaaS) provider platform is placed between the supply chain parties and the IOTA Tangle. The industrial-scale miners, logistic companies and processors publish their data to the platform which creates a record within their cloud and simultaneously creating the transaction on the Tangle. In this setup, it is also easier for the customers and other participants to gather information about the origin trail. They just have to type in the ID created by the mining site on a user-friendly created web front end.

After mining the ore, at the processing step, the information systems of the processing machinery and employees scan the QR code when they receive the ore. They now check if the received material was signed by the address of the mining team. If the attached QR code was not stored in the Tangle or with a unique identifier (X) which does not have information stored about the origin on their legal concessions, they will not accept the ore for further processing. In addition, they are not allowed to accept any material that does not have any kind of identifier on them. This can also be traced back if they process ore which was not scanned previously. Therefore, also risk factor 3 is addressed. The person that wants to fake ore origin on the mine site needs to get access to the node which signs the transactions with drill hole information inside. The person would have to create fake transactions and create labels on its own which he then sticks onto the artisanally mined ore. This seems rather unlikely as also the algorithm which creates new identifiers after specific intervals (time or tonnage) needs to be bypassed.

Risk factor 2 and 3: Artisanal mined ore gets mixed with industrial-scale mined ore in the transportation process

This step is especially risky since physical goods, as well as information systems, are interacting at an intra-company level. At this point, also because the transported material is high in value, the mining company needs to attach beacons with trinary hardware on the packages. The contract between miner and logistic providers includes a list of the created IDs X from the mining site. The logistics provider also takes note of rough estimates of weight and volume.

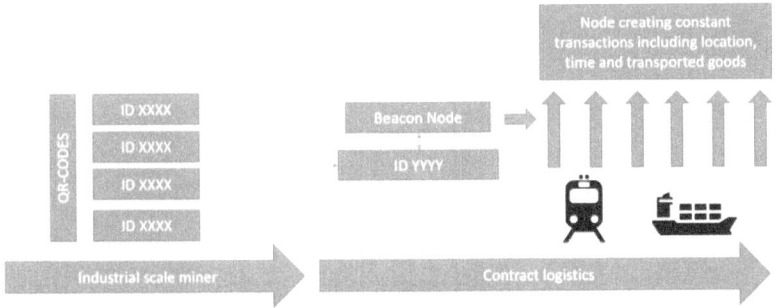

Figure 32: Information flow between mining and logistics company – aggregating n ID X into ID Y (Own representation)

The idea is that the contract logistics company is creating a new identifier Y which is attached to a transaction aggregating all X IDs from the contract and also the information about weight and volume. The new transaction is now constantly published to the Tangle by the trinary hardware chip running on the battery-powered beacon along the transportation path. As analyzed earlier even if e.g. the ship between Africa and China loses connection to the internet, a local Tangle network on the boat itself can later be reattached to the main Tangle. As a result, the Tangle now constantly holds records of where ID Y was located at what time. ID Y of course still contains information about the origin of the mined ore (X IDs), volume and weight.

Risk factor 4: Processors buy artisanally mined ore

The unethical sources can enter the supply chain here in two ways. If the logistics company needs to repack or relabel the transported goods or tries to mix in other sources, the receiving party has to demand proof of the Y ID. Although, as mentioned earlier the logistics company has no benefit to ship the wrong goods to the successor.

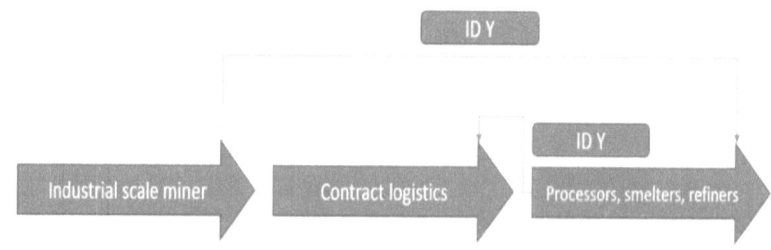

Figure 33: Double checking of ID Y to mitigate risk of mixed goods (Own representation)

Since the miner does have a direct contract with the processing company, they are able to exchange the Y ID before. Since the Y ID includes every information needed to ensure ethically sourced metal, the supply chain is safe if the processor only accepts goods that were included in the transaction that is assigned to ID Y. If the processor company decides to buy artisanal mined ore and process it further, they will have a hard time to proof the origin to the next party, the battery component producers of the downstream processes. They require full proof of the origin of the material they buy, which is easy to document if the correct Y IDs can be validated. It is nearly impossible at this stage to fake an entire supply chain for a single party. For the further steps in the supply chain, it is important that an aggregated information package summarized into a transaction is constantly published to the Tangle. To demonstrate this, an example information flow is shown in figure 34.

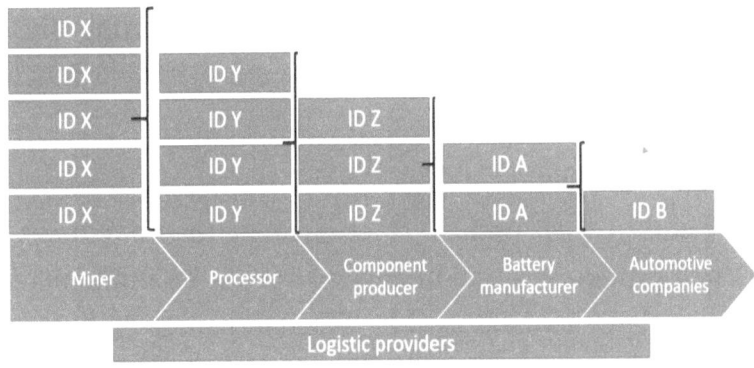

Figure 34: Information flow based on subsequent Tangle transactions along the supply chain (Own representation)

The last ID B which represents an individual car is leading to a transaction that contains all IDs of ID As. Those ID As lead to transactions that are including all ID Zs. This step can be iterated to the beginning.

Therefore, every final product (car) can be tracked down via the Tangle back through the whole supply chain. Every time a node was publishing a transaction, information about IDs, location and time stamp were deposited on the Tangle. It has to be noted that all those transactions cannot be changed by any party afterwards and are accessible to everyone. In reality, this concept will create thousands of transactions per second due to the high complexity and involved parties within global automotive supply chains. Here the benefits of the Tangle due to its unlimited scalability, quantum security and feeless structure show why this technology is such a great fit. Since IOTAs Tangle is specially made for the IoT environment it is also easy to automate tasks between parties, e.g. a harbor terminal scans the transport ships over beacon / RFID technology and automatically forwards the information if the shipment is valid or not to the next party. This automation can also lead to reducing the risks significantly. If a chemical processing plant is automatically validating the inserted ore and attaches a new ID to the output units, there is no chance of any human misbehavior.

To give the whole Tangle-based tracing system even more security and also an option for companies to pro-actively market their ethic products, an audit system should be implemented. The International Organization for Standardization (ISO) and the International Automotive Task Force (IATF) have identified changes in automotive supply chains and, in response, have released the new industry standards ISO9001:2015 and IATF16949:2016. These norms include adjusted standards for the automobile industry and their suppliers in order to increase the transparency and quality of automotive supply chains. To confirm that a company meets the required industry standards, such as the IATF16949:2016 and ISO9001:2015, certifications are performed. This is usually done by third party audits with an external auditor (Brauweiler, 2015, p. 7).

A proposal to the certification bodies would be to include the requirement for a tracing system in the supply chain into the norms. Companies who want to be certified to their standards then need to prove that they utilize an efficient tracing system. Two types of audits are recommended for an efficient setup. The first audit should be at the beginning of the certification, where the auditor needs to investigate if the nodes and systems are set up properly at every physical site. The second audit which needs to be within a recurring interval, e.g. every 6 months, needs to check on mine sites, processors and logistics providers if they utilize the tracing concept properly. Another important topic is the sharing of trainings and lessons learned between all parties.

The application of the Tangle-based tracing concept which was developed in chapter 3 has shown that the Tangle is a good match to the problems automotive supply chains face nowadays. Especially the immutable, scalable and feeless nature makes it an efficient tool to provide more guidance of origin of battery components. A summary of all the research questions and a further outlook will be given in the next chapter.

5-CONCLUSION AND OUTLOOK

The environment for automotive companies and their supply chains has changed significantly in the past years. Globalized value chains with multiple production sites over the world create new challenges for the companies. With the hyperbolic market share rise of electronic vehicles, the topic of transparency and supplier ethics becomes more and more important. This is because of the cobalt and copper supply markets. Both materials are needed to produce the batteries of the vehicles, are high in demand and lack in available supply. The possibility of being able to dig for those metals by hand, combined with high metal prices makes it especially interesting for artisanal mining in undeveloped countries. The DRC, which produces a major part of the world's cobalt supply still relies by a significant share on those methods.

Simultaneously, advancing technologies such as DLTs, are presenting new challenges but also opportunities for supply chains, requiring new solutions and methods to ensure the success of a business. DLTs became public because of data laws and a general skepticism regarding security. In addition, data scandals from big corporations harmed the trust given to data origin and credibility. The main goal of this thesis was to answer three research questions. The first question asked what potentials do DLTs have in general and what the reason for that is. In the following, the question is answered and reasons are given. With a new type of structure such as decentralized networks, these technologies claim to improve security and decrease dependency from third parties.

The literature review shows that the area of DLTs is very broad and complex, but most projects are based on blockchain technology. Blockchain technologies require miner pools to run their transaction network. This leads to the transactions having high fees, take a long time to process and also need a lot of energy which harms the environment. Due to the aggregation of mining pools and not being quantum-proof, the blockchain protocol also brings security breaches with it.

Since this setup brings also severe disadvantages in the use cases automotive companies need, especially in terms of being scalable, tamper-proof and immutable, the research on blockchain was extended to the latest evolutions. The concept of a DAG-based concept running on the Tangle by the IOTA Foundation was found out to be the most suitable to address the problems evaluated in the introduction. The IOTA Tangle is a technology that goes beyond blockchain and is managed by the IOTA Foundation in Berlin. They do not only develop the software algorithms of the Tangle but moreover, they also created JINN labs which investigates on a trinary hardware processor that leverages the usability of the nodes. The second research question asked by this research work was how the IOTA Tangle differs from blockchain solutions and what makes its design unique and what the advantages are. Due to the different setup, running a DAG, the technology allows suspending miners. Therefore, the network participants themselves need to approve the transactions of other users. This eliminates the need for transaction fees, miner pools and creates a highly scalable solution. Special features of the Tangle such as partitioning or being able to run side Tangles make the concept an even better fit to the tracing problem of electronic vehicle supply chains.

Chapter 3.3 then focused on existing supply chain tracing systems. To sum up, existing tracing systems are made up of multiple soft- and hardware technologies that collect data and store them within e.g. the cloud of a company. A major problem is that the single steps are not integrated from end-to-end and data breaches between steps lead to losing transparency (black box). Most of the existing concepts utilize RFID, QR-Codes and Beacons on the hardware side. The research shows that Bluetooth Beacons are most effective but also are the most expensive and therefore it was detected that a mix of different options adapted to a specific situation is the best approach.

For the reason that the sum of possible variants and supply chain routes is too extensive, a real-world example of an electronic vehicle supply chain was worked through in chapter 4.1. This was the starting point for the Tangle-based tracing concept provided in the next chapter. The literature shows that upstream processes are more focused in one area, mostly the DRC, with only a small amount of parties included. In addition, the second half of the electronic vehicle supply chain tends to be in more stable and secure jurisdictions. Two points of failure can be extracted out of the findings. First, we have the point where artisanal miners sell their mined ore to the traders. The next step includes the traders buying such sources, combining them with legally mined ore and selling them further to the next party. This shows the importance to solve the problems of transparency, already at the beginning of the supply chain. The concept developed on the Tangle should ensure that only industrial-scale produced material is entering the supply chain in the first place and not just trace back if existing batteries contain unethical sources.

The final research question posed, what could be an effective process for companies to trace electronic vehicle supply chains utilizing the IOTA Tangle to solve the problems of artisanal mining. The thesis combined the literature review, with IOTA's latest developments and established a tracing concept in chapter 4.2. The recommendation is to give every item mined by the company its unique identifier which gets populated along the supply chain. Every party needs to run nodes that create transactions every time they forward the goods to the next party. On the receiving processes, they need to check the provided information with the records stored on the Tangle before further progressing. Therefore, every final product (car) can be traced down via the Tangle back through the whole supply chain. Every time a node was publishing a transaction information about IDs, location and time stamp were deposited on the Tangle.

To give the whole Tangle-based tracing system even more security and also an option for companies to pro-actively market their ethic products, an audit system should be implemented. The International Organization for Standardization (ISO) and the International Automotive Task Force (IATF) have identified changes in automotive supply chains and, in response, have released their new industry standards. The requirement of a tracing system that is audited on a recurring interval should be implemented into those standards and should be mandatory to obtain such a certification document.

As IOTA and the Tangle are still in development mode and updates enter the market nearly every month it should be investigated further how their approach works in the future. Especially the tip selection algorithm, permanodes, hashing algorithms and side-Tangles open up for further research. One aspect that is criticized for DLTs, in general, is the data quality. Since the technology is only the enabler to store, transfer and secure the data, the data generation itself needs to be of high quality and accuracy.

Companies should not wait for the technologies to become available to the market, they should proactively take place in the development and accompany in use case creation. This is what makes the IOTA Tangle so special, there are hundreds of renowned companies working together with the foundation to build a M2M payment solution for the future.

One of the latest achievements of the foundation is a cooperation with the Object Management Group (OMG) which is responsible for creating standards for companies worldwide. They are well known for their standards Business Process Model and Notation (BPMN) and Unified Modeling Language (UML). They plan to release an IOTA standard together with the ISO at the end of 2020 (Soley, 2019, p. 1).

The foundations have been laid for IOTA to become the industry standard for IoT M2M payments and to supersede blockchains by going beyond their vision, and it will be interesting to see what the future brings to this project.

LIST OF REFERENCES

Agence France-Presse (2019): DR Congo army moves to secure Chinese mine, https://www.theeastafrican.co.ke/news/africa/DR-Congo-deploys-army-to-Chinese-mine/4552902-5161972-ex6ry4/index.html (03.08.2019).

Ahn, In-Kyoon/Kwak, Seung Hyun/Lee, Seung Young (2018): Audit Report on Congo Dongfang International Mining sarl. prepared for LG Chem, Korea

Amnesty International Ltd (2016): "THIS IS WHAT WE DIE FOR" HUMAN RIGHTS ABUSES IN THE DEMOCRATIC REPUBLIC OF THE CONGO POWER THE GLOBAL TRADE IN COBALT, Ausgabe 2/2010 in Policy Sciences, London

Azevedo, Marcelo/Campagnol, Nicolo/Hagenbruch, Toralf/Hoffman, Ken/Lala Ajay, Ramsbottom, Oliver (2018): McKinsey&Company – Lithium and Cobalt – a tale of two commodities, London

Banchirigah, Sadia Mohammed/Hilson, Gavin (2010): De-agrarianization, re-agrarianization and local economic development: Re-orientating livelihoods in African artisanal mining communities, Guildford

BBC News (2019): Ford accelerates electric vehicle investment, https://www.bbc.com/news/business-47644323 (14.06.2019)

Eylert, Bernd/Mohnke, Janett (2012): Signaturverfahren in Sicherheit in der Informationstechnik, Berlin

BGR (2017): Cobalt from the DRC – Potential, Risks and Significance for the Global Cobalt Market, Hannover

Blockchain, 2019: Size of the Bitcoin blockchain from 2010 to 2019, by quarter (in megabytes), https://www-statista-com.gw.akad-d.de/statistics/647523/ worldwide-bitcoin-blockchain-size/ (27.05.2019)

Blockchain.com (2019a): Average Confirmation Time, https://www.blockchain. com/de/charts/avg-confirmation-time?timespan=all (18.06.2019)

Blockchain.com (2019b): Hashrate Verteilung Eine Abschätzung der Hashrate-Verteilung unter den größten Mining-Pools, https://www.blockchain .com/pools (18.06.2019)

Brauweiler, Jana/Zenker-Hoffmann, Anke/Will, Markus (2015): Auditierung und Zertifizierung von Managementsystemen, Wiesbaden

Brennan, W. John/Barder, Timothy E. (2016): Arthur D Little - Battery Electric Vehicles vs. Internal Combustion Engine Vehicles - A United States-Based Comprehensive Assessment, Boston

Burgwinkel, Daniel (2016): Blockchain Technology – Einführung für Business- und IT Manager, Berlin

Buterin, Vitalik (2014): On Mining, https://blog.ethereum.org/ 2014/06/19/mining/ (15.08.2019)

Chand, Manish/Ramachandran, Navin/ Stoyanov, Danail (2018): Techniques in Coloproctology - Robotics, artificial intelligence and distributed ledgers, in: surgery: data is key! Volume 22, issue 9, p. 645-648

Courtois, Nicolas T./Bahack, Lear (2014): On Subversive Miner Strategies and Block Withholding Attack in Bitcoin Digital Currency, New York

De Bock, Joris (2018): E-Mobility Business Models – Making The Change Happen, Berlin

Dehman, Nabylah Abo (2018): DRILLING DOWN INTO THE COBALT SUPPLY CHAIN: HOW INVESTORS CAN PROMOTE RESPONSIBLE SOURCING PRACTICES, London

Derks, Jona/Gordijn, Jaap/Siegmann, Arjen (2018): Electron Markets - From chaining blocks to breaking even: A study on the profitability of bitcoin mining from 2012 to 2016, Heidelberg

Eyal, Ittay/Sirer, Emin Gün (2018): Majority Is Not Enough: Bitcoin Mining Is Vulnerable in Communications of the ACM, New York

Frankel, Todd C./Robinson, Michael/Ribas, Jorge (2016): The cobalt pipeline: Tracing the path from deadly hand-dug mines in Congo to consumers' phones and laptops, Washington Post

Gaudlitz, Eva (2016): Indoor Tracking Using Beacons or RFID – What Are the Differences?, https://www.infsoft.com/blog-en/articleid/60/indoor-tracking-using-beacons-or-rfid-what-are-the-differences (15.07.2019)

Gayvoronskaya, Tatiana/ Meinel, Christoph/ Schnjakin, Maxim (2018): Blockchain – Hype oder Innovation Technischer Bericht Nr. 113, URL: https://publishup.uni-potsdam.de/opus4ubp/frontdoor/deliver/index/docId/ 10314/file/ tbhpi113.pdf (25.09.2019)

Glencore plc (2017): Annual Report, https://www.glencore.com/dam/ jcr:62bed41c-1627-4bf5-bc43-

cf5518ba1193/glen-2017-annual-report.pdf (15.06.2019)

Handelsblatt (2019): Amazon und GM planen Beteiligung an E-Pickup-Hersteller,https://www.handelsblatt.com/unternehmen/industrie/rivian-amazo n-und-gm-planen-beteiligung-an-e-pickup-hersteller/23980304.html?ticket =ST-63374861-5uw50ju79ufU1nPiR4D3-ap5 (15.07.2019)

Hein, Cathrin/Wellbrock, Wanja (2019): Rechtliche Herausforderungen von Blockchain-Anwendungen – Straf-, Datenschutz-, und Zivilrecht, Wiesbaden

Heinz, Arnold (2019): VW baut eigene Batteriezellenfertigung – 1 Mrd. Euro bereitgestellt, https://www.elektroniknet.de/markt-technik/automotive/vw-baut-eigene-batteriezellenfertigung-165349.html (22.07.2019)

Helmar (2018): When would it be required to intentionally partition the Iota transaction graph? ,https://iota.stackexchange.com/questions/1973/when-wo uld-it-be-required-to-intentionally-partition-the-iota-transaction-graph (18.05.2019)

Hentschel, Thomas/Hruschka, Felix/Priester, Michael (2003): Artisanal and Small-Scale Mining - Challenges and Opportunities, London

Higer, Rene (2019): Alles über IOTA: einfach und verständlich, https://www.iota-wiki.com/de/#JINN (16.07.2019)

Hocking, Mathew/Kan, James/Terry, Chris/Begleiter, David (2016): Deutsche Bank Markets Research - Welcome to the Lithium-ion Age, Sydney

Coinmetrics (2019): Network Data Charts - Fees, transaction,

mean USD,
https://coinmetrics.io/charts/#assets=btc_log=false_roll=14_right=
averageFeeUsd_zoom=1470301200000,1558742400000
(17.06.2019)

Hume, Neil (2019): Glencore shares fall after DRC mine collapse,
https://www.ft.com/content/542c9768-98ec-11e9-8cfb-
30c211dcd229 (27.07.2019)

Institute for Economics & Peace (2019): Global Peace Index 2019:
Measuring Peace in a Complex World, Sydney,
http://visionofhumanity.org/reports (15.04.2019)

IOTA Foundation (2019): Coordicide - The next step in IOTA's
Evolution, https://www.youtube.com/watch?v=guNNqEeu6gY
(12.07.2019)

Ivancheglo, Sergey (2018): Economic Clustering and IOTA,
https://medium.com/@comefrombeyond/economic-clustering-and-
iota-d3a77388900, 02.07.2019

Jentzsch, Nicola (2011) : Datenskandale und Super-GAUs: Neues
Denken im Datenschutz notwendig?, in: Deutsches Institut für
Wirtschaftsforschung (DIW), Berlin, Vol. 78, Iss. 22, pp. 20

Kamoto Copper Company S.A. (KCC) (2018): KATANGA
MINING LIMITED NI 43-101 TECHNICAL REPORT ON THE
MATERIAL ASSETS OF KATANGA MINING LIMITED,
LUALABA PROVINCE, DEMOCRATIC REPUBLIC OF
CONGO,
http://www.katangamining.com/~/media/Files/K/Katanga-mining-
v2/ operations/reportsoperational/technical-report-march-2018.pdf
(15.06.2019)

Leong, Christine/Viskin, Tal/Stewart, Robyn (2018): Accenture –

Tracing The Supply Chain, How blockchain can enable traceability in the food industry, New York

Maize (2018): LONG STORY SHORT - WHAT IS IOTA?, https://www.maize.io/en/content/ what-is-iota (06.06.2019)

Malone, David & O'Dwyer, K.J. (2014): Bitcoin Mining and its Energy Footprint, Maynooth

Petroff, Alana (2018): Carmakers and big tech struggle to keep batteries free from child labor, https://money.cnn.com/2018/05/01/technology /cobalt-congo-child-labor-car-smartphone-batteries/index.html (30.05.2019)

Randall, Tom (2017): The Electric-Car Boom Is So Real Even Oil Companies Say It's Coming, https://www.bloomberg.com/news/articles/ 2017-04-25/electric-car-boom-seen-triggering-peak-oil-demand-in-2030s (14.06.2019)

Rosenberger, Patrick (2018): Bitcoin und Blockchain – Vom Scheitern einer Ideologie und dem Erfolg einer revolutionären Technik, Münster

Schaller, Robert R. (1997): Moore's law: past, present, and future, in: IEEE Spectrum USA Piscataway, Volume 34, Issue 6,

Schiener, Dominik (2017): A Primer on IOTA (with Presentation), https://blog.iota.org/a-primer-on-iota-with-presentation-e0a6eb2cc621 (02.04.2019)

Schiener, Dominik (2018): Videoclip: Bosch Software Innovations (User): Insights from the expert: IOTA for Supply Chain, https://www.youtube.com/ watch?time_continue=65&v=cZ_q7QRwHcM (15.08.2019). Used part in 01:02

Smith, Brett (2018): How Cobalt is Used in Lithium-Ion Batteries, https://www.azomining.com/Article.aspx?ArticleID=1421, (28.07.2019)

Smits, Adams (2018): Kalongwe: Building a significant ASX-listed mining company in the heart of the world's richest copper-cobalt belt – Company presentation, https://nzuricopper.com.au/wp-content/uploads/2018/05/22May_Nzuri_RSS_Corporate-AS-5pm-23-5-18-FINAL-print-version.pdf (15.06.2019)

Soley, Richard (2019): Insights 3 – Standardization, https://www.youtube.com/watch?v=ZqZTHSchOEI&feature=youtu.be (07.08.2019)

Sønstebø, David (2016): Honest Data – Ensuring Data Integrity, https://blog.iota.org/honest-data-f4e25bdac5ad (02.07.2019)

Sønstebø, David (2017): IOTA Development Roadmap, https://blog.iota.org/iota-development-roadmap-74741f37ed01 (02.07.2019)

Spengler Arnim., Malkwitz A., Ehlers J., Thesing A. (2017): Innovative Produkte und Dienstleistungen in der Mobilität - Supply Chain Tracking im BIM Modell, Wiesbaden

Swanson, Tim (2015): Consensus-as-a-service: a brief report on the emergence of permissioned Singapore, distributed ledger systems, http://www.ofnumbers.com/wp-content/uploads/2015/04/Permissioned-distributed-ledgers.pdf (15.06.2019)

Tesla Inc. (2018a): Tesla's revenue from FY 2008 to FY 2018 (in million U.S. dollars), https://www-statista-com.gw.akad-d.de/statistics/272120/revenue-of-tesla/ (22.05.2019)

Tesla Inc. (2018b): **Tesla's net income/loss attributable to common stockholders from FY 2015 to FY 2018 (in million U.S. dollars)**, https://www-statista-com.gw.akad-d.de/statistics/272130/net-loss-of-tesla/ (22.05.2019)

Tonks, Alex/Rodriguez, Daniel (2018): CRU Perth Tech Metals Briefing

Tran, Tam/Luong, Van-Tri (2015): Lithium Process Chemistry - Lithium Production Processes, South Korea

Truong, Nguyen/Jayasinghe, Upul/Um, Tai-Won/Lee, Gyu Myoung (2016): A Survey on Trust Computation in the Internet of Things, in: THE JOURNAL OF KOREAN INSTITUTE OF COMMUNICATIONS AND INFORMATION SCIENCES (J-KICS). Volume 33, Issue 10-27

Tsurukawa, Nicolas/Prakash, Siddharth/ Manhart, Andreas (2011): Social impacts of artisanal cobalt mining in Katanga, Democratic Republic of Congo, Freiburg

Uddin, Md. Akther (2018): Bitcoin: A hype or digital gold! The Financial Express, Chittagong

US Geological Survey (2019): Average cobalt spot price in the United States from 2014 to 2018 (in U.S. dollars per pound), https://www-statista-com.gw.akad-d.de/statistics/339743/average-spot-price-of-cobalt-in-the-us/ (22.05.2019)

Verband der Automobilindustrie (2019): Mittelstand, https://www.vda.de/de/ themen/automobilindustrie-und-maerkte/mittelstand/zulieferindustrie-und-mittelstand.html (22.07.2019)

Vlassenroot, Koen/Van Bockstael Steven (2008): Artisanal Diamond Mining: Perspectives and Challenges, Belgium

Wannenwetsch, Helmut (2010): Integrierte Materialwirtschaft und Logistik - Beschaffung, Logistik, Materialwirtschaft und Produktion, Berlin

Waters, Richard (2019): IBM unveils first standalone quantum computer, https://www.ft.com/content/f0d63c74-12c6-11e9-a581-4ff78404524e, 02.07.2019

Weik, Martin (2000): Computer Science and Communications Dictionary – scalability, Springer, Boston

West, James (2011): Decreasing Metal Ore Grades: Are They Really Being Driven by the Depletion of High-Grade Deposits?, in: Journal of Industrial Ecology, Volume 15, issue 1, p. 1-16

Whitley, Angus/Futonaka, Mei (2019): Toyota Strikes Deal With World's Top Supplier of Electric Car Batteries, https://www.bloomberg.com/news /articles/2019-07-17/toyota-strikes-battery-deal-with-catl-in-push-for-electric-cars (28.07.2019)

DECLARATION

I hereby declare and confirm that this thesis is entirely the result of my own original work. Where other sources of information have been used, they have been indicated as such and properly acknowledged. I further declare that this or similar work has not been submitted for credit elsewhere.

18.11.2019_____

Date and location of signing   Student´s signature

Munich, 18.11.2019